I0755816

THE

CHEMISTRY OF COMMON LIFE.

BY

JAMES F. JOHNSTON, M.A., F.R.S., F.G.S.,

ETC., ETC.,

AUTHOR OF "LECTURES ON AGRICULTURAL CHEMISTRY AND GEOLOGY," "A CATECHISM OF AGRICULTURAL CHEMISTRY AND GEOLOGY," ETC.

ILLUSTRATED WITH NUMEROUS WOOD ENGRAVINGS.

VOL. I.

FOURTH EDITION.

NEW YORK:

D. APPLETON AND COMPANY,

346 & 348 BROADWAY.

M.DCCC.LV.

DEDICATION.

TO SIR DAVID BREWSTER,

K.H. D.C.L. F.R.S. V.P.R.S., EDINBURGH, ASSOCIATE OF THE INSTITUTE OF FRANCE, AND PRINCIPAL OF ST. LEONARD'S COLLEGE, ST. ANDREW'S.

MY DEAR SIR DAVID,

I dedicate this little Work to you, partly that I may have the honour of associating with it a name so eminent in science as yours, and partly for the opportunity it gives me of expressing my sense of the many obligations I owe you as an old and tried friend.

Being yourself not only a lover and assiduous cultivator of science, but a remarkable extender of its boundaries—a leader in one of its most interesting and intricate departments—and an anxious diffuser of the results of general scientific research—I am certain of your sympathy in the following attempt to render popular some of the more immediately applicable results of that branch to which I have myself been now long devoted. If we, whose profession it is to follow the progress of science, can scarcely keep pace with the advance of our several departments, it must be especially necessary, from time to time, to present its more striking novelties, in an intelligible form, to the general public.

With sincere wishes that your health may be long preserved, and that optical science may still for many years number you among its most illustrious cultivators,

Believe me,

MY DEAR SIR DAVID,

Your obliged friend,

JAMES F. W. JOHNSTON.

DURHAM, *October*, 1853.

INTRODUCTION.

The common life of man is full of wonders, Chemical and Physiological. Most of us pass through this life without seeing or being sensible of them, though every day our existence and our comforts ought to recall them to our minds. One main cause of this is, that our schools tell us nothing about them—do not teach those parts of modern learning which would fit us for seeing them. What most concerns the things that daily occupy our attention and cares, are in early life almost sedulously kept from our knowledge. Those who learn any thing regarding them, must subsequently teach themselves through the help of the press: hence the necessity for a Popular Chemical Literature.

It is with a view to meet this want of the Public, and at the same time to supply a Manual for the Schools, that the present Work has been projected. It treats, in what appears to be their natural order, of THE AIR WE BREATHE and THE WATER WE DRINK, in their relations to human life and health—THE SOIL WE CULTIVATE and THE PLANT WE REAR, as the sources from which the chief sustenance of all life is obtained—THE BREAD WE EAT and THE BEEF WE COOK, as the representatives of the two grand divisons of human food—THE BEVERAGES WE INFUSE, from which so much of the comfort of modern life, both savage and civilised, is derived—THE SWEETS WE EXTRACT, the history of which presents so striking an illustration of the economical value of chemical science—THE LIQUORS WE FERMENT, so different from the sweets in their action on the system, and yet so closely connected with them in chemical history—THE NARCOTICS WE INDULGE IN, as presenting us with an aspect of the human constitution which, both chemically and physiologically, is more mysterious and wonderful than any other we are yet acquainted with—THE ODOURS WE ENJOY, and THE SMELLS WE DISLIKE; the former because of the beautiful illustration they present of the recent progress of organic chemistry in its relations to the comforts of common life, and the latter because of their intimate connection with our most important sanitary arrangements—WHAT WE BREATHE FOR and WHY WE DIGEST, as relating to functions of the body at once the most important to life, and the most purely chemical in their nature—THE BODY WE CHERISH, as presenting many striking phenomena, and performing many interesting chemical functions not touched upon in the discussion of the preceding topics—and lastly, the CIRCULATION OF MATTER, as exhibiting in one view the end, purpose, and method of all the changes in the natural body, in organic nature, and in the mineral kingdom, which are connected with and determine the existence of life.

It has been the object of the Author in this Work, to exhibit the present condition of chemical knowledge, and of matured scientific opinion upon the subjects to which it is devoted. The reader will not be surprised, therefore, should he find in it some things which differ from what is to be found in other popular works already in his hands or on the shelves of his library.

CONTENTS OF VOL. I.

CHAPTER I.

THE AIR WE BREATHE.

Height of the earth's atmosphere; it is one of the elements of the ancients.—Composition of the atmosphere.—Oxygen, preparation and properties of.—Nitrogen, preparation and properties of.—Proportions of these elements in the air; their adaptation in kind and quantity to the existing condition of things.—Uses of the oxygen and nitrogen.—Uses of the carbonic acid; its importance to vegetable life.—Deleterious influence upon animal life.—The "Poison Valley" of Java.—Importance of the watery vapour of the air; its constant circulation.—Formation of rain and dew; their many uses.—Accidental constituents of the air; ozone, nitric acid, and ammonia.—Vapours which rise from the surface of the earth, and saline matters from the sea.

The earth we inhabit is surrounded by an atmosphere of air, the height of which is known to be at least forty-five miles. It presses upon the earth with a weight equal at the level of the sea to about 15 lb. on every square inch of surface. As we ascend high mountains, this weight becomes less; and as we go down into deep mines, it becomes sensibly greater.

We breathe this atmospheric air, and without it we could not live a single moment. It floats around the earth in almost perpetual motion; and according to the swiftness with which it moves, it produces gentle breezes, swift winds, or terrible tornadoes.

Though very familiar to us, and regarded with little

curiosity, this air is yet very wonderful, both in itself and in its uses. Imperfect as the knowledge of the ancients was, they recognised its importance by giving it a place among what they regarded as the four primal elements of nature—fire, air, earth, and water.

Yet, though apparently pure and elementary, it is by no means either a simple or pure substance. It is a mixture of several different kinds of matter, each of which performs a beautiful and wise part in relation to animal and vegetable life. Four substances, at least, are known to be necessary to its composition. Two of these, oxygen and nitrogen, form nearly its entire bulk; the two others, carbonic acid and watery vapour, being present only in minute quantities.

Oxygen is a kind of air or gas, which, like the atmosphere itself, is without colour, taste, or smell. A candle burns in it with much greater brilliancy and rapidity than in common air. Animals also breathe in it with an increase of pleasure; but it excites them, quickens their circulation, throws them into a state of fever, and finally kills them, by excess of excitement. They live too rapidly in pure oxygen gas, and burn away in it like the fast-flaring candle.

Fig. 1.

This gas is easily prepared by mixing the chlorate of potash of the shops with a little sand, powdered glass, or oxide of manganese, and heating the mixture in a flask over a spirit-lamp. When it melts, the gas is given off, and will soon fill the flask. It cannot be seen by the eye, or detected by any of the other senses. Its presence may be readily shown, however, by introducing a lighted taper or a bit of red-hot charcoal, or of kindled phosphorus at the end of a wire (fig. 1). The brilliancy of the burning will prove the presence of the gas.

Nitrogen is also a kind of air which, like oxygen, is void of colour, taste, and smell; but a lighted candle is instantly extinguished, and animals cease to breathe when introduced into it. We obtain this gas by putting a bit of phosphorus into a small cup over water, kindling it, and inverting over it a bottle, dipping with its mouth into the water (fig. 2). When the phosphorus has ceased to burn, and the bottle has become cool, it may be corked and removed from the water. If a lighted taper be now introduced into the bottle, it will immediately be extinguished, showing that only nitrogen remains (fig. 3). In this process, the burning phosphorus removes the oxygen from the air contained in the bottle, and leaves only the nitrogen.

Fig. 2.

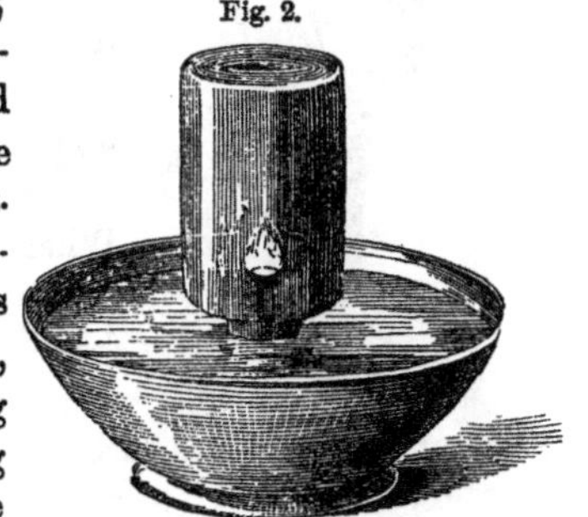

Fig. 3.

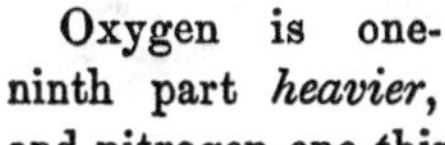

Oxygen is one-ninth part *heavier*, and nitrogen one thirty-sixth part *lighter* than common air.

Carbonic acid is a kind of air which, like oxygen and nitrogen, is void of colour; but, unlike them, possesses a slight odour, and a perceptibly sour taste. Burning bodies are extinguished, and animals cease to breathe when introduced into it. It is one-half heavier than common air, and can therefore be poured through the air from one vessel to another (fig. 4). When passed through lime-water,* it makes it milky (fig. 5), forming with the dissolved lime an insoluble white powder, which, because it contains carbonic acid, is called *carbonate* of lime, and is the same thing as chalk. It is the escape of this gas which gives their sparkling briskness to fermented liquors, to soda-water, and to the waters of some mineral springs.

Fig. 4.

Fig. 5.

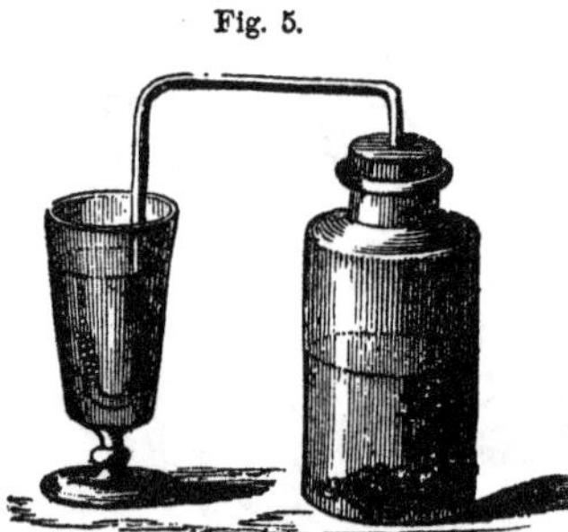

Carbonic acid is easily prepared by pouring vinegar upon common soda, or diluted spirit of salt (muriatic acid) upon chalk or limestone. The gas rises in bubbles through the liquid, and, in consequence of its weight, remains in the lower part of the vessel. As it collects it gradually ascends, driving the common air before it, and at

* Lime-water is formed by pouring water upon slaked lime, shaking them well together, and allowing the mixture to settle. The clear liquid contains a portion of the lime in solution, and is therefore called *lime*-water.

last flows, as water would do, over the edge of the vessel. Its rise may be shown by introducing two lighted tapers, as in the figure (fig. 6), when the lower one will be seen to go out, while the upper one is still burning.

Fig. 6.

By *watery vapour* is meant the steam or vapour visible, or invisible, which ascends from a surface of water when exposed to the air. When water is spilt upon the ground in dry weather, it soon disappears : it rises in invisible vapour, and floats buoyantly among the other constituents of the atmosphere.

These four substances the air every where and always contains. They are all necessary to the daily wants of animal and vegetable life; but the two gases, oxygen and nitrogen, form so large a proportion of the whole that we are accustomed to say of dry air, that it consists of nitrogen and oxygen only, in the proportion of 4 gallons of the former to 1 of the latter. More correctly, however, air, when deprived of the watery vapour and carbonic acid it contains, consists, in 100 gallons, of 79 of nitrogen mixed with 21 of oxygen; or of—

	By measure.
Nitrogen, ...	79
Oxygen, ...	21
	100

The carbonic acid exists in the air in very small proportion. At ordinary elevations there are only about 2 gallons of this gas in every 5000 of air—$\frac{1}{2500}$th part of the whole. It increases, however, as we ascend, so that at heights of 8000 or 10,000 feet the proportion of carbonic acid is nearly doubled. Even this increased quantity is very small; and yet its presence is essential to the existence of vegetable life on the surface of the earth.

But being heavier than common air, it appears singular that the proportion of this gas should increase as we ascend into the atmosphere. Its natural tendency would seem to be rather to sink towards the earth, and there to form a layer of deadly air, in which neither animal nor plant could live. But independent of winds and aërial currents, which tend to mix and blend together the different gases of which the air consists, all gases, by a law of nature, tend to diffuse themselves through each other, and to intermix more or less speedily, even where the utmost stillness prevails and no wind agitates them. Hence a light gas like hydrogen does not rise wholly to the utmost regions of the air, there to float on the heavier gases; nor does a heavy gas like carbonic acid sink down so as to rest permanently beneath the lighter gases. On the contrary, all slowly intermix, become interfused, and mutually intercorporated, so that the hydrogen, the carbonic acid, and the other gases which are produced in nature, may be found everywhere through the whole mass, and a comparatively homogeneous mixture uniformly overspreads the whole earth. In obedience to this law, carbonic acid in all places slowly rises or slowly sinks, as the case may be, and thus, on the whole, a uniform purity is maintained in the air we breathe. If it seems to linger in sheltered hollows like the deadly gas-lake of Java, it is because the fatal air issues from the earth as rapidly as it can diffuse itself upwards through the atmosphere; and if it rest more abundantly on the mountain top, it is because the leaves of plants, and the waters of the sea, absorb it from the lower layers of the air faster than it can descend to supply their demands.

The watery vapour varies in quantity with the climate and temperature of the place. It is less in cold seasons and climates generally than in such as are hot. It seldom forms more than $\frac{1}{60}$th, or less than $\frac{1}{2000}$th of the bulk of the air.

The presence of carbonic acid in the atmosphere is shown by the formation of a white film of carbonate of lime on the surface of lime-water when this is exposed to the air. The presence of watery vapour may be shown on the hottest days by pouring ice-cold water into a tumbler or water-bottle, when the vapour of the air will rapidly condense on the outer surface of the vessel in the form of drops of dew.

The purposes which we know to be served by these several constituents of the atmosphere show both that they are all essential to the composition of the air, and that in quantity as well as kind they have been beneficently adjusted to the composition, the wants, and the functions of animals and of plants.

Thus, as to the oxygen—

From every breath of air which the animal draws into its lungs it extracts a quantity of oxygen. The oxygen thus obtained is a part of the natural food of the animal, which it can obtain from no other natural source, and new supplies of which are necessary to it every moment. The oxygen of the atmosphere, therefore, is essential to the very existence of life in the higher orders of animals.

The candle burns also, and all combustible bodies kindle in the air, only because it contains oxygen. This gas is a kind of necessary food to flaming and burning bodies; so that were it absent from the earth's atmosphere, neither light nor heat could be produced from coal, wood, or other combustible substances.

But the proportion, also, in which oxygen exists in the air is adjusted to the existing condition of things. Did the atmosphere consist of oxygen only the lives of animals would be of most brief duration, and bodies once set on fire would burn so fast as to be absolutely beyond control. The oxygen is therefore mixed with a large proportion of nitrogen. This gas, not being poisonous, as carbonic acid is, harmlessly dilutes the too active oxygen. It weakens and prolongs its action on the system as water dilutes wine or spirits, and assuages their too fiery influence upon the animal frame.

Then, as to the carbonic acid—

Every green leaf that waves on field or tree sucks in, during the sunshine, this gas from the air. It is as indispensable to the life of the plant as oxygen is to the life of the animal. Remove carbonic acid from the air and all vegetable growth would cease. It must, therefore, be a necessary constituent of the atmosphere of our earth.

But carbonic acid is poisonous to animals. It is for this reason that the proportion of this gas contained in the air is so very small. Were this proportion much greater than it is, animals, as they are now constituted, could not breathe the atmosphere without injury to their health.* On the

* The most remarkable natural example of an atmosphere overloaded with carbonic acid gas is the famous Poison Valley in the island of Java, which is thus described by an eyewitness :—

"We took with us two dogs and some fowls to try experiments in this poisonous hollow. On arriving at the foot of the mountain we dismounted and scrambled up the side about a quarter of a mile, holding on by the branches of trees. When within a few yards of the valley we experienced a strong nauseous suffocating smell, but on coming close to its edge this disagreeable odour left us. The valley appeared to be about half a mile in circumference, oval, and the depth from thirty to thirty-five feet; the bottom quite flat; no vegetation; strewed with some very large (apparently) river stones; and the whole covered with the skeletons of human beings, tigers, pigs, deer, peacocks, and all sorts of birds. We could not perceive any vapour or any opening in the ground, which last appeared to us to be of a hard sandy substance. It was now proposed by one of the party to enter the valley; but at the spot where we were this was difficult at least for me, as one false step would have brought us to eternity, seeing no assistance could be given. We lighted our cigars, and, with the assistance of a bamboo, we went down within eighteen feet of the bottom. Here we did not experience any difficulty in breathing, but an offensive nauseous smell annoyed us. We now fastened a dog to the end of a bamboo eighteen feet long, and sent him in: we had our watches in our hands, and in fourteen seconds he fell on his back, did not move his limbs or look round, but continued to breathe eighteen minutes. We then sent in another, or rather he got other hand, that growing plants may be able to obtain a sufficiently large and rapid supply of carbonic acid from a gaseous mixture which contains so little, they are made to hang out their many waving leaves into the atmosphere. Over the surface of these leaves are sprinkled countless pores or mouths, which are continually employed in separating and drinking in carbonic acid gas. The millions of leaves which a single tree spreads out, and the constant renewal of the moving air in which they are suspended, enable the living plant to draw an abundant supply for all its wants from an atmosphere already adjusted to the constitution of living animals.*

This constant action of the leaves of plants is one of the natural agencies by which the proportion of carbonic acid in the lower regions of the atmosphere is rendered less than it is in the higher regions.

So, also, the watery vapour of the atmosphere is not less necessary to the maintenance of life. The living plant consists of water to the amount of nearly three-fourths of its whole weight, and from the surface of its leaves water is continually rising into the air in the form of invisible vapour.

Were the air absolutely dry, it would cause this water to evaporate from their leaves more rapidly than it could be supplied to them by the soil and roots. Thus they would

loose, and walked in to where the other dog was lying. He then stood quite still, and in ten minutes fell on his face, and never afterwards moved his limbs: he continued to breathe seven minutes. We now tried a fowl, which died in a minute and a half. We threw in another, which died before touching the ground. During these experiments we experienced a heavy shower of rain; but we were so interested by the awful sight before us that we did not care for getting wet. On the opposite side, near a large stone, was the skeleton of a human being, who must have perished on his back, with his right hand under his head. From being exposed to the weather the bones were bleached as white as ivory. I was anxious to procure this skeleton, but any attempt to get it would have been madness."—LOUDON.

* A common lilac-tree with a million of leaves, has about four hundred thousand millions of pores or mouths at work, sucking in carbonic acid; and on a single oak tree, as many as seven millions of leaves have been counted.

speedily become flaccid, and the whole plant would droop, wither, and die.

The living animal in like manner is made up for the most part of water. A man of 154 lb. weight contains 116 lb. of water, and only 38 lb. of dry matter. From his skin and from his lungs water is continually evaporating. Were the air around him perfectly dry his skin would become parched and shrivelled, and thirst would oppress his feverish frame The air which he breathes from his lungs is loaded with moisture. Were that which he draws in entirely free from watery vapour, he would soon breathe out the fluids which fill up his tissues, and would dry up into a withered and ghastly mummy. It is because the simoom and other hot winds of the desert approach to this state of dryness, that they are so fatal to those who travel on the arid waste.

Thus the moisture which the atmosphere contains is also essential to the maintenance of the present condition, both of animal and vegetable life: it pervades the leaves and pores of plants, and finds admission to the lungs and general system of animals.

There are, besides, other beautiful purposes which this moisture serves. When the summer sun has sunk beneath the horizon, and coolness revisits the scorched plant and soil, the grateful dew descends along with it and moistens alike the green leaf and the thirsty land—the invisible moisture of the air thickens into hazy mists, and settles in tiny pearls on every cool thing. How thankful for this nightly dew has nature everywhere and always appeared, and how have poets in every age sung of its beauty and beneficence!

Let us attend for a moment to the cause of this descent of the dew, and to the way in which it seems to select, as it were, the spots on which it will fall.

All bodies on the surface of the earth radiate, or throw out rays of heat in straight lines—every warmer body to every colder—and the whole earth itself is continually sending rays of heat upwards through the clear air into free cold space. Thus on the earth's surface all bodies strive, as it were, after an equality of temperature (an equilibrium of heat), while the surface as a whole tends gradually towards a cooler state. But while the sun shines on any spot this cooling will not take place, for the surface there receives for the time more heat than it gives off; and, when the sun goes down, if the clear sky be shut out by a canopy of clouds, these will arrest and again throw back to the earth a portion of the heat which escapes by radiation, and will thus prevent it from being dissipated. At night, then, when the sun is absent, the earth will cool the most—on clear nights also more than when it is cloudy; and when clouds only partially obscure the sky, those parts will become coolest which look towards the clearest portions of the heavens.

Again, the quantity of vapour which the air is capable of holding in suspension is dependent upon its temperature. At high temperatures, in warm climates, or in warm weather, it can sustain more—at low temperatures, or in cold weather, less. Hence, when a current of comparatively warm air, loaded with moisture, ascends to, or comes in contact with, a cold mountain-top, it is cooled down, is rendered incapable of holding the whole of the vapour in suspension, and therefore leaves behind, in the form of a mist or cloud encapping the lofty summit, a portion of its watery burden. The aqueous particles which float in this mist appear again on the plains below, in the form of streams or springs, which bring nourishment at once, and a grateful relief to the thirsty soil.

So, when the surface cools by radiation, the air in contact with it must cool also; and, like the warm currents on the mountain side, must forsake a portion of the watery vapour it has hitherto retained. This water, like the floating mist on the hills, descends in particles almost infinitely minute. These particles collect on every leaflet, and suspend them selves from every blade of grass in drops of "pearly dew."

And mark here a beautiful adaptation. Different substances are endowed with the property of radiating their heat, and of thus becoming cool with different degrees of rapidity. Those substances which in the air become cool first must also attract first, and most abundantly, the particles of falling dew. Thus, in the cool of a summer's evening the grass-plot is wet, while the gravel-walk is dry; and the thirsty pasture and every green leaf are drinking in the descending moisture, while the naked land and the barren highway are still unconscious of its fall.

And from the same atmospheric store of watery vapour come the refreshing showers which descend in our temperate zone, and the rushing rains which fall in torrents within the tropical regions—only the mode in which they are made to descend is somewhat different.

In the upper regions of the atmosphere currents of cold air are continually rushing from the north, and currents of warm air from the south. When two such currents of unequal temperature, each loaded with moisture, meet in the atmosphere, they mix, and the mixture has the mean temperature of the two; but air of this mean temperature is incapable of holding in suspension the mean quantity of watery vapour contained in the two currents. Hence, as on the mountain side, a cloud is formed, and the excess of moisture collecting into drops, falls to the earth in the form of rain.

When we consider how small a proportion of watery vapour exists in the air—that were it all to come down at once over the whole earth, it would cover the surface only to a depth of 5 inches—we cannot think without amazement of the vast and continuous effects it produces. The quantity of rain which falls yearly on our islands would cover them, were it all to fall at once, to a depth of from 25 to 30 inches; and, except the table-land of central Spain, there are few places in western Europe where the depth of yearly rain is less than 20 inches. And all this rain descends from an atmosphere which does not contain more, probably, at any one time, than falls yearly in dew alone over the whole earth.*

In descending, also, this rain discharges another office: it washes the air as it passes through it, dissolving and carrying down those accidental vapours which, though unwholesome to man, are yet fitted to assist the growth of plants. It thus ministers in another double manner to our health and comfort, purifying the air we breathe, and feeding the plants on which we live.

As soon, again, as the rain ceases to fall, and the clear sky permits the sun's rays once more to warm the surface of the earth, vapours begin to rise anew, and the sweeping winds dry up the rains and dews from its moistened surface. There are regions of the globe, also, where unending summer plays on the surface of the wide seas, and causes a perpetual evaporation to lift up unceasing supplies of water into the air. These supplies the wind wafts to other regions; and thus the water which descends in rain or dew in one spot, is replaced by that which mounts up in vapour from another. And all this to maintain unbroken that nice adjustment which fits the constitution of the atmosphere to the wants of living things!

How beautiful is the arrangement by which water is thus constantly evaporated or distilled, as it were, into the atmosphere—more largely from some, more sparingly from

* How, among the hills in tropical countries, the rain really rushes down may be inferred from the fact, that among the Khassaya hills, north of Calcutta, the yearly fall of rain amounts to 610 inches (50 feet), of which 550 fall in the six rainy months, beginning in May. As much as 25½ inches have been observed to fall in a single day.

other spots—then diffused equally through the wide and restless air, and afterwards precipitated again in refreshing showers which cleanse the tainted air, or in long-mysterious dews. But how much more beautiful the contrivance—I might almost say the instinctive tendency—by which the dew selects the objects on which it delights to fall; descending first on every living plant, copiously ministering to the wants of each, and expending its superfluity only on the unproductive waste!

And equally kind and beautiful, when understood, nature is seen to be in all her operations. Neither skill nor materials are ever wasted; and yet she ungrudgingly dispenses her favours apparently without measure, and has subjected dead matter to laws which compel it to minister, and yet with a most ready willingness, to the wants and comforts of every living thing.

Four substances, therefore—oxygen, nitrogen, carbonic acid, and watery vapour—are essential to the composition of the atmosphere, and they are adjusted, both in kind and quantity, to the existing condition of things. But besides these, the air contains also many other substances in minute and indefinite proportions. Of these, some are formed in the air itself, some rise in vapour from the surface of the earth, and some ascend from the waters of the sea.

Of those which are formed in the air itself, two are deserving of especial mention—ozone, and nitric acid.

The former of these is merely oxygen gas in what is called a more exalted chemical condition than that in which it usually exists. Into this condition it is brought by the action of the sun's rays, of electricity, and of many other agencies. In this form it acts upon and combines more readily with all other substances. Among the other useful purposes it is supposed to serve, I mention the oxidation*

* When a substance combines with oxygen, it is said to be *oxidised*, or to undergo *oxidation*.

of the organic, often noxious, substances which rise into the atmosphere, and of those vegetable and other compounds in the soil, upon which depend its general fertility, and the abundant production of the food of plants.

Ozone is probably never absent from the atmosphere; but it is always present in a proportion too minute to admit of being determined either by weight or by measure. It is more abundant in winter, on the tops of mountains, and after a storm has purified the air. It is probably more serviceable to us than we are yet aware of.

Nitric acid, the other important substance I have mentioned as being formed in the air, is probably more abundant than ozone. It is commonly known by the name of aquafortis, and consists of nitrogen and oxygen only—the two main constituents of the atmosphere. Every flash of lightning which darts across the sky, and every electric spark, great or small, which in any other form passes through the air, causes a minute proportion of these two gases, along the line of its course, to unite together and produce nitric acid. And as this passage of electricity through the air is frequent almost everywhere, and in the tropical regions is distinctly visible nearly every day of the year, I am inclined to regard this acid as a constant constituent of atmospheric air. Whether it is essential or indispensable to the present condition of things, we have not as yet the means of determining; but it has been ascertained by actual experiment that this acid is at least very frequently present in the air, even of European countries, and falling rain is sometimes actually sour from the quantity of nitric acid it contains. This acid is very favourable to vegetable growth—and is, indeed, one of the substances which the falling rains and dews are appointed to wash out of the air, and in doing so to bring down to plants a valuable form of food, which is thus daily prepared for them among the winds of heaven.

From the surface of the earth, again, there arise continually into the air vapours and gases of various kinds. The vegetable and animal bodies which undergo decay in manifold circumstances, and the numerous substances which are burned in the air, all produce chemical compounds, which, being volatile or gaseous, ascend and mingle with the atmosphere. Some of these, like ammonia and sulphuretted hydrogen, are perceptible to the smell, while others are altogether inappreciable by the senses. The steaming marsh also, beneath the summer's sun, sends forth fatal miasms which prostrate the body in fever, though neither the senses can perceive, nor our more refined chemical tests as yet detect their presence; living volcanoes likewise belch forth their vapours; and a thousand chemical operations, natural and artificial, pour out their fetid streams and volatile exhalations. All these ascend from the earth, are caught by the winds, wafted more or less speedily from their birthplace, and mingle with the general air. Thus the atmosphere must contain accidental substances almost without end, which are not essential to its constitution, and which rise into the aërial sea because of their lightness, just as liquid impurities spontaneously flow, or solid impurities are washed down by the rivers into the waters of the great ocean.

Of these substances which thus ascend from the earth in the form of gas, ammonia deserves especial notice, because of the important function which some agricultural writers have ascribed to it in reference to vegetable growth. This gas, which is familiar to every one in the smell of common hartshorn,* is formed during the putrefaction of animal and vegetable substances in the presence of water and air, and is the principal cause of the smell which heaps of such putrefy-

* The liquid hartshorn of the shops is only water impregnated with the gas ammonia.

ing matters give off. It is continually rising, therefore, into the atmosphere from many parts of the earth's surface. It has consequently been found in very minute quantity in the air, wherever it has been sought for. Some, therefore, deem it an essential constituent of our air. In this respect, however, it must be distinguished from nitric acid, which we know to be produced in the atmosphere itself by purely physical causes, and to be altogether independent of the previous existence of life. It is possible, as I have elsewhere shown,* that ammonia may be so produced also; in which case we might not only acknowledge it for an essential constituent of the atmosphere, but discover in its existence, and constant reproduction there, a wise provision for the maintenance of vegetable growth.

Further, from the ever-moving sea, the winds which raise it into rolling waves, and lash it into foam, sweep upwards the light spray, and mingle it with the rushing air. Thus, far inland and over high mountains, the salty particles are carried, and all the contents of sea water are mingled with the universal atmosphere. Hence the host of foreign substances which must float around us, commingled with those which we know to be absolutely necessary to the maintenance of animal and vegetable life, is almost inconceivable.

The accumulation of all these foreign matters in the air would, in course of time, render it unwholesome to animal life—perhaps unfit for the healthy development even of vegetable forms. But the waters of heaven, as I have described, ascend and descend continually to wash and purify it. They serve as a natural conservative check.

Thus simple as the air appears, its scientific history as a whole is somewhat complicated. The adjustment of its constituents involves many interesting particulars, and the ar-

* *Lectures on Agricultural Chemistry and Geology*, second edition, p. 288.

rangements by which the constant presence of its essential constituents is secured, both in kind and quantity, are very numerous; yet we cannot fail to perceive both a physical beauty, and a wise contrivance in them all.

CHAPTER II.

THE WATER WE DRINK.

Importance of water in nature.—Composition of water.—Hydrogen gas; how prepared; the lightest of known substances, and an inflammable gas; exists in nearly all combustible substances; is always converted into water when these substances are burned.—In water hydrogen is combined with oxygen.—What is meant by a chemical combination.—Water without taste and smell; importance of this.—Cooling property of water.—Relation of water to other liquids.—It dissolves many solid substances; hence natural water never pure.—Quantity of mineral matter in some known river, spring, and sea waters.—Composition of the solid matter in sea water; in the Thames water at Kew; and in that of the Kent Water Company.—Lime held in solution in water by carbonic acid.—Why calcareous waters incrust their channels, petrify, and deposit sediments in boilers.—Impurity of spring waters in large towns, about farmhouses, and near graveyards.—Composition of well water from Highgate Hill.—Well waters in the *dunes* of Bordeaux; their analogy to the waters of Marah.—Water absorbs its own bulk of carbonic acid at all pressures.—How this explains the liveliness of champagne and soda-water, the bursting of bottles, the briskness and deadness of beer, &c.—Excess of oxygen in the air contained in water; importance of this to the lives of fishes.—More oxygen near the surface of the sea.—Why air obtained from snow contains less oxygen.

THE water we drink is next in importance to the air we breathe. It forms three-fourths of the weight of living animals and plants, is the most abundant substance we meet with on the face of the earth, and covers, to an unknown depth, at least three-fourths of its entire surface.

Pure water consists of two simple or elementary substances,* oxygen and hydrogen. The former of these exists

* By *simple* or *elementary* substances, chemists understand such as cannot by any known means be resolved or split up into more than one: sulphur, phosphorus, gold, silver, iron, &c., are examples of such simple substances.

also in common air, and has been described in the previous chapter.

Hydrogen is a kind of air or gas which, when pure, is without colour, taste, or smell. It differs, however, from all the three gases (oxygen, nitrogen, and carbonic acid) described in the preceding chapter; *first*, in being the lightest of all known substances; and, *second*, in taking fire, and burning in the air when a lighted taper is brought near it.

Fig. 7.

It is readily prepared by putting a few pieces of metallic zinc or iron into a bottle or flask, and pouring over them a quantity of oil of vitriol (sulphuric acid) diluted with twice its weight of water. When a sufficient quantity of the gas has been produced to drive out the common air from the bottle, a gas jet-burner, or a bit of glass tube, or of a tobacco pipe thrust through a cork, may be put into the mouth of the bottle, when a jet of gas will issue which may be lighted by a taper. It burns with a very pale flame. When a perfectly dry, cool, glass tumbler or bottle is held over the flame (fig. 7), dew will be seen to condense on the inner side of the glass, which will gradually collect into little visible globules, and will finally trickle down in the form of drops of pure water. This water is formed by the burning of the hydrogen from the bottle in the oxygen of the air. During this burning it *combines* with the oxygen, and water is produced. The extreme lightness of the hydrogen may be shown by extinguishing the gas, and causing it to ascend into a small empty balloon placed over the jet* (fig. 8). When the balloon is full of gas it will readily ascend, showing not only that the hydrogen is lighter than common air, but that it is so much lighter as to be able to raise heavy bodies through the air along with it. It is to the lightness of this gas that we owe the power of travelling through the air in ordinary balloons.

Fig. 8.

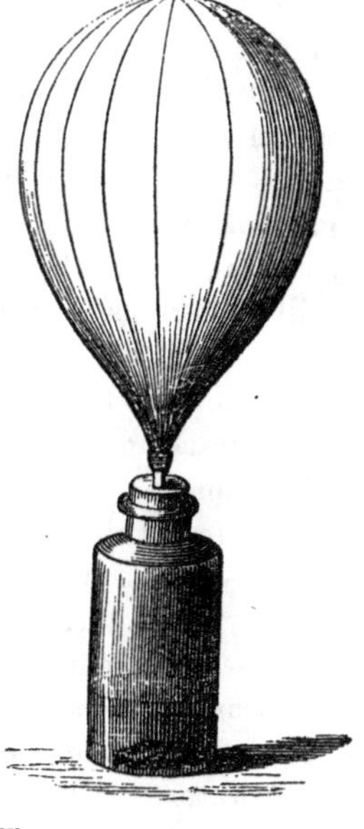

Hydrogen exists in a great many other substances besides water—in bituminous coal, in wood, in oils and fats, in coal gas, and in nearly all combustible substances; but whenever it is completely burned in the air, water is formed by its union with oxygen, as in the burning of the simple jet above described. Thus, in nearly all cases of combustion, water is one of the substances produced, though it generally rises into the air in the form of invisible vapour.

Water thus formed consists of oxygen and hydrogen, in the proportions by weight of—

			Per cent.
Oxygen,	8	or	88.88
Hydrogen,	1	„	11.11
	9	„	100

—or every 9 lb. of pure water contain 8 lb. of oxygen and 1 lb. of hydrogen.

In atmospheric air, as we have seen, there are at least

* Such little balloons, made of thin membranes, are sold by the opticians.

four substances present which are essential to its existence But between air and water there is this important chemical distinction, that in the former the constituents are merely mixed together, while in the latter they are *chemically combined.* When nitrogen and oxygen are *mixed* together to form common air, each of them retains its gaseous form, and all its properties unaltered; but when hydrogen and oxygen are *combined* to form water, they severally lose both their original gaseous form, and all their distinctive properties, both physical and chemical. Water is not light, like hydrogen, nor will it burn as that gas does; neither will bodies burn in it as they do so readily and brilliantly in oxygen gas.

Now, when bodies combine chemically, they always form a new substance different in its properties from those which have been employed in producing it; and, indeed, it is one of the wonders which modern chemistry has made known to us, that hydrogen, which burns so readily, should form so large a part of water, our great extinguisher of flame; and that oxygen, so indispensable to animal life, should form eight-ninths of a liquid in which few terrestrial animals can live for more than three or four seconds of time.

That water is indispensable to animal and vegetable life, appears both from its forming so large a proportion of the bodies of living animals and plants, and from some other considerations which have been stated in the preceding chapter. But many of the properties which water possesses are wonderfully conducive to our comfort, to the supply of our daily wants, and to the maintenance of the existing condition of things.

1°. Thus, even the unheeded property of its freedom from smell and taste is important to animal comfort. Sweet odours are grateful to our nostrils at times, and pleasant savours give a relish to our rarer kinds of food. But health fails in an atmosphere which is ever loaded with incense and perfumes, or where the palate is daily pampered with high-seasoned dishes and constant sweets. The nerves of smell and taste do not bear patiently a constant irritation, and the whole body suffers when a single nerve is continually jarred. Hence it is that water and air, which have to enter so often into the animal body, and to penetrate to its most delicate and most sensitive organs and tissues, are made so destitute of sensible properties that they can come and go to any part of the frame without being perceived. Noiselessly, as it were, they glide over the most touchy nerves; and, so long as they are tolerably pure, they may make a thousand visits to the extremest parts of the body without producing the most momentary irritation or sense of pain. Externally, also, they can be applied to the most delicate, inflamed, or skinless parts of the body, not only without irritating, but generally with the most grateful and soothing effects. These negative properties, which are common both to air and water—though, as I have said, they are rarely thought of—are nevertheless most essential to our daily comfort.

2°. Again, water possesses a cooling property, which is very grateful to all living things. The priceless value of water in "a dry and thirsty land" arises mainly from the necessity of constantly supplying that which, in a dry and warm atmosphere, is constantly evaporating from the skin and the lungs. But in all climates water has a cooling power, which gives it a new value to the hot and fevered animal. When taken into the mouth and stomach, or when poured over the inflamed skin, it cools more than an equal weight of any other liquid or solid substance we could apply. This arises from the circumstance, that it takes more heat to give a sensible warmth to water than to an equal weight of any other common substance. Thus the same quantity of heat which is required to raise the temperature of 1 lb. of water a single degree (from 60° to 61° for example), would give an equal increase of temperature to 30 lb. of quicksilver; and so, again, to convert water into vapour, requires more heat than an equal weight of any other liquid we consume. Hence, when water evaporates from the skin, it serves as a constant cooler of the surface; while the vapour, which escapes with the breath, cools equally the interior of the body. It is really very interesting to observe how the great capacity of liquid water for heat makes it so gratefully cooling as it enters the body; and how its still greater capacity for heat, when passing from the liquid state to the state of steam, enables it so constantly to bear away from us the germs of fever, as it escapes from our bodies in the form of insensible vapour.

3°. But the peculiar composition of water is also a very important circumstance to animal and vegetable life. It consists of oxygen and hydrogen; and all the solid parts of animals and plants contain these same elements in large proportion. In the dry wood of the tree, for example, and in the dry flesh and bone of the animal, both are present. Now, as the plant and animal increase in size, oxygen and hydrogen are required for the formation of their growing parts, and water is everywhere at hand to supply these necessary ingredients. This is a chemical duty which no other liquid but water could equally perform. Water, in discharging this duty, is not merely the drink, as we usually call it, but is really part of the food both of animal and plant.

4°. Further, pure water possesses the property of mixing with some other fluids, such as alcohol (strong spirits) in all proportions, merely weakening or diluting their strength. With others, again—as with oil—it refuses to mingle. Solid substances it has the property of dissolving; and upon this property depend many of the most useful purposes served by water, in reference both to animal and vegetable life.

If a piece of sugar and a piece of glass be put together into a quantity of water, the former will dissolve and disappear, while the water will remain for any length of time in the water unaltered in form or in weight. Water does not dissolve all bodies therefore. Sugar is soluble—glass is insoluble in this liquid.

Again, if into two equal quantities of water we introduce loaf-sugar and common salt—the sugar into the one and the salt into the other—as long as they are respectively dissolved and disappear, we shall see that 1 lb. of water will dissolve perhaps 2 lb. of sugar, forming a thick syrup, while it will only dissolve 5¾ oz. of common salt. Thus, of those substances which dissolve in water, some are much more soluble—disappear, that is, in larger quantity than others do.

In nature, water is never found perfectly pure; that which descends in rain is contaminated by the impurities it washes out of the air; that which rises in springs, by the substances it meets with in the earth itself. In rivers, the impurity of the water is frequently visible to the eye. It is often of a red colour as it flows through rocks of red marl which contain much oxide of iron in their composition; it descends milky from the glaciers of Iceland and the slopes of the Andes, because of the white earth it holds in suspension; it is often grey or brown in our muddiest English rivers; it is always brown where it issues from boggy lakes, or runs across a peaty country; it is sometimes black to the eye when the quantity of vegetable matter is excessive, as in the Rio Negro of South America; and it is green in the Geysers of Iceland, in the Swiss lakes, among the islands of the South Sea, and around our own islands, because of the yellow matters which it everywhere holds in suspension or solution. Only in clear and deep waters—like those of the Bay of Naples, and in parts of the Pacific, where minute objects may be seen on the bottom some hundreds of feet down—is the real blue colour natural to water, in large masses, distinctly perceptible.*

But among the rocky and other materials which water meets with in and upon the earth, there are many which it can dissolve, as it does salt and sugar, and the presence of which cannot be detected by the sense of sight. Hence the clearest and brightest of waters—those of springs and transparent rivers, even when filtered—are never pure; they all contain in solution a greater or less quantity of saline matter, sometimes so much as to give them a decided taste, and to form what are hence called *mineral* waters.

Among the purest natural waters hitherto examined is that of the Loka, in the north of Sweden, which flows over hard impenetrable granite and other rocks, upon which water produces little impression. It contains only $\frac{1}{20}$ of a grain (0.0566) of solid mineral matter in the imperial gallon. Some waters in the granite regions of the north of Scotland, and even some springs which rise through the green-sand in Surrey, contain as little as 4 or 5 grains in the gallon. The water which is supplied to the city of Edinburgh contains 7 to 14 grains in the gallon,† and that of the Thames, near London, about 21. These are both comparatively pure waters, and are very good for general consumption. That of the river Wear, which supplies the city of Durham, contains 15½ grains in the gallon, and is still a good water for domestic use. That which is used in the town of Sunderland, and is obtained from the lower new red sandstone, contains 27 grains in the gallon. Some of the other waters supplied to and used in London and its neighbourhood,

* This is the blue which is seen in the azure grotto of the Isle of Capri, in the Bay of Naples, and in the deep, indigo-like waters of some parts of the Mediterranean and Adriatic seas.

† This is 1 to 2 parts by weight in 10,000 of the waters—a gallon of pure water at 60° Fahr. weighing 70,000 grains.

and which are not derived from the Thames, contain, in a gallon—

New River Company, . .	19¼	grs. in the gallon.
East London Water Company, .	23¼	" "
Kent Water Company, . .	29¾	" "
Hampstead Water Company, .	35½ to 40	" "
Deep-bore wells, . . .	33 to 38	" "

Other drinking-waters contain more even than these. Some which are in constant use contain twice as much—even the waters of the holy Jordan contain 73 grains to the gallon—but generally, in the waters of average purity which are employed for domestic purposes, there are not present more than from 20 to 30 grains of solid matter in the imperial gallon.

Generally speaking also, rain water which falls in remote country districts is the purest; then comes river water; next, the water of lakes; after these, common spring-waters; and then the water of mineral springs. The waters of the Black Sea, and the Sea of Azof, which are only brackish, follow next; then those of the great ocean; then those of the Mediterranean,* and inland sea; and last of all come those of lakes which, like the Caspian Sea, the Dead Sea, and Lake Aral, possess no known outlet. All the solid matter which the rivers carry into the sea remains there, while the water which brings it is continually rising again in vapor. This vapor, as we have seen, descends in the form of rain on the interior of continents, and there dissolves, and thence carries down new supplies of mineral matter to the sea. In this way saline matter has accumulated in the ocean till its waters have become briny and bitter to the taste. In the same way, also, it has accumulated in the Caspian and Dead Seas—the more rapid evaporation

* Off the coast of Havre, for example, the sea water does not contain more than 3¼; while in the Mediterranean it contains 3¾ per cent of saline matter. The composition or quality of this saline matter is nearly the same in each case.

in these parts of the world, the unfrequent rains, and probably the neighborhood of deposits of rock-salt, having aided in making these inland waters so much salter than those of the great oceans. The waters of the great ocean, and its branches, contain from 2200 to 2800 grains of saline matter in the gallon; those of the Dead Sea in some places 11,000; in others, as much as 21,000 grains, or one-fourth part of their whole weight. Those of a small lake east of the steppes of the Wolga, contain as much as three-fifths of their weight of saline matter.

Common salt is the most abundant kind of saline matter which occurs in sea water; but it contains also the chlorides of calcium and magnesium,* and some other salts, in considerable proportion. One of the most recent examinations of sea water has been made by Riegel. His sample, taken off the coast of Havre, contained, in 1000 parts by weight $31\frac{1}{2}$ parts of solid matter (2250 grains in the gallon), consisting of—

Chloride of sodium (common salt),	24.632
Chloride of potassium,	0.307
Chloride of calcium,	0.439
Chloride of magnesium,	2.564
Bromide of magnesium,	0.147
Sulphate of lime (gypsum),	1.097
Sulphate of magnesia † (Epsom salts),	2.146
Carbonate of lime (chalk),	0.176
Carbonate of magnesia,	0.078
	31.586

The reader will observe that, next to common salt, the compounds of magnesia are most abundant in sea water. The same is the case with the waters of the Dead Sea and

* *Chlorine* is a greenish-yellow gas, which combines with metals and forms *chlorides; bromine*, a dark red liquid, forms *bromides; iodine*, a lead-grey solid, forms *iodides*.

† Sulphuric acid, or oil of vitriol, unites with lime, magnesia, soda, &c., and forms *sulphates*.

other very salt lakes, and to this they chiefly owe their acrid bitter taste.

Besides the substances above named, traces of phosphate of lime, of silica, of the oxides of iron and manganese, of iodine, of fluorine, and even of lead, copper, silver, and arsenic, have been detected in sea water. Indeed, we know that, being the common reservoir into which all soluble substances are washed down by the rains and rivers, we ought to find in the sea traces of all the soluble substances which are capable of existing together in the same solution.

Even the spring and river waters employed for domestic purposes often contain a considerable variety of substances. Thus the water of the Thames, taken at Kew by the Grand Junction Water Company, and that supplied to London by the Kent Water Company, contain, respectively, in an imperial gallon—

	Thames water.		Kent Water Company.	
Carbonate of lime (chalk),	10.90	grs.	7.02	grs.
Sulphate of lime (gypsum),	3.26	„	11.03	„
Nitrate of lime,	trace	„	0.07	„
Carbonate of magnesia,	1.17	„	3.42	„
Chloride of sodium (common salt),	1.40	„	3.50	„
Sulphate of soda,	0.18	„	—	„
Chloride of potassium	—	„	0.44	„
Sulphate of potash	0.61	„	0.70	„
Silica	0.44	„	0.76	„
Iron, alumina, and phosphates,	0.67	„	trace	„
Organic matter, with a trace of ammonia	3.07	„	2.61	„
	21.70	„	29.55	„

Lime, in combination with carbonic acid (carbonate), and with sulphuric acid (sulphate), is the most abundant substance in these two waters. Indeed, it very often exists in large quantity, especially in spring waters; and it is chiefly to the lime and magnesia they contain, that what are called *hard* waters owe their property of curdling with soap.

Pure waters are always soft; and when a water is tolerably soft, it may be inferred that it does not contain any large proportion of lime, or magnesia.

Waters which contain much lime are often bright and sparkling to the eye, and agreeably sweet to the taste. They generally become somewhat milky when boiled, and leave a sediment, which incrusts the inside of kettles or boilers. When strongly impregnated with lime, they will even deposit a calcareous coating along their channels as they flow in the open air, or will incrust, or petrify, as it is called, any solid substances which are immersed in them. These circumstances are owing to the peculiar way in which the lime is held in solution.

We have already seen that, if a current of carbonic acid be made to pass through lime-water (as in fig. 5), the transparent liquid will become at first milky, from the formation of carbonate of lime, which remains suspended in the form of a very fine powder; but if the current of carbonic acid be continued, the milkiness will gradually disappear, the carbonate of lime will be re-dissolved, and the liquid will again become clear. The carbonate of lime is held in solution by an excess of carbonic acid.

If, now, the clear solution be poured from one vessel to another for a number of times, it will gradually give off this excess of carbonic acid into the air, and become milky again. This is what happens when calcareous springs incrust the sides of their channels, as in Auvergne, or at Matlock and Knaresborough in our own country. Or if a coin or other solid substance be introduced into the solution, bubbles of carbonic acid gas will gradually be given off, and the substance will become incrusted with lime—the carbonate of lime which falls. This is exactly what takes place in a petrifying well. Or if the solution be heated over the fire, the excess of carbonic acid is driven off, the solution becomes milky as before, and the whole of the lime falls in the form of carbonate, leaving the water nearly pure. The incrustation in our kettles and boilers is chiefly produced in this latter way. Hard waters, therefore, are generally made much softer and purer by boiling. Should much of the lime, however—as in the water supplied by the Kent Water Company, above noticed—be in the state of gypsum, mere boiling will not alone soften it; but if a little soda be added to it during the boiling, this will separate the lime of the gypsum also.

As this solvent power of water enables it to take up many substances from the rocks and soils through which it passes, it often happens that, in the neighbourhood of dwellings and farmyards, and especially in towns, the water of wells becomes very impure, and even unwholesome to drink. The rains that fall upon the filth that accumulates in towns wash out the soluble substances it contains, carry them into the soil, and through this, by degrees, to the wells by which the wants of the inhabitants are supplied. This has often been productive of serious and fatal disease. It shows, therefore, the propriety of preventing, as far as possible, the accumulation of refuse, and, where such accumulation is unavoidable, of placing it at the greatest distance from wells which yield water for daily use. And, especially, it shows the necessity of bringing water from a distance for the supply of large cities.

The neighbourhood of grave-yards is equally fitted, with the accumulation of town refuse, to adulterate water with undesirable admixtures. The water of a well which is close to the old churchyard on the top of Highgate Hill, has lately been examined by Mr. Noad, and found to contain as much as 100 grains of solid matter to the gallon, consisting of—

Nitrate of lime, ...	40.12 grains.
Nitrate of magnesia, ...	17.06 „
Sulphate of potash, ...	17.04 „
Sulphate of soda (Glauber salts), . .	9.52 „
Chloride of sodium (common salt), . .	9.63 „
Chloride of calcium. . . .	5.91 „
Silica, ...	0.90 „
	100.18 grains.

This large amount of *nitrates** is traced to the neighbouring grave-yard, as such compounds are generally produced where animal matters decay in porous soils. While the buried bodies were more recent, animal matters of a more disagreeable kind would probably have been found in the well, as I have myself found them in the water of wells situated in the neighbourhood of farmyards.

Well-waters sometimes contain vegetable substances also of a peculiar kind, which render them unwholesome, even over large tracts of country. In sandy districts the decaying vegetable matters of the surface-soil are observed to sink down and form an ochrey *pan*, or thin yellow layer in the subsoil, which is impervious to water, and through which, therefore, the rains cannot pass. Being arrested by this pan, the rain water, while it rests upon it, dissolves a certain portion of the vegetable matter; and when collected into wells, is often dark coloured, marshy in taste and smell, and unwholesome to drink. When boiled, the organic matter coagulates, and when the water cools separates in flocks, leaving the water wholesome, and nearly free from taste or smell. The same purification takes place when the water is filtered through charcoal, or when *chips of oak wood are put into it.* These properties of being coagulated by boiling, and by the tannin of oak wood, show that the organic matter

* The *nitrates* consist of nitric acid (aquafortis) combined with lime, magnesia, &c. Saltpetre is *nitrate of potash*, consisting of nitric acid combined with potash, and so on.

contained in the water is of an albuminous character, or resembles white of egg. As it coagulates, it not only falls itself, but it carries other impurities along with it, and thus purifies the water—in the same way as the white of egg clarifies wines and other liquors to which it is added.

Such is the character of the waters in common use in the *Landes* of the Gironde around Bordeaux,* and in many other sandy districts. The waters of rivers, and of marshy and swampy places, often contain a similar coagulable substance. Hence the waters of the Seine at Paris are clarified by introducing a morsel of alum, and the river and marshy waters of India by the use of the nuts of the *Strychnos potatorum*, of which travellers often carry a supply. One or two of these nuts, rubbed to powder on the side of the earthen vessel into which the water is to be poured, soon causes the impurities to subside. In Egypt, the muddy water of the Nile is clarified by rubbing bitter almonds on the sides of the water-vessel in the same way.

In all these instances the principle of the clarification is the same. The albuminous matter is coagulated by what is added to the water, and in coagulating it embraces the other impurities of the water, and carries them down along with it.

These cases, and especially that of the sandy Landes of Bordeaux, and elsewhere, throw an interesting light upon the history of the waters of Marah, as given in the fifteenth chapter of Exodus.

"So Moses brought Israel from the Red Sea; and they went out into the wilderness of Shur; and they went three days in the wilderness and found no water. And when they came to Marah, they could not drink of the waters of Marah, for they were bitter: therefore the name of it was called Marah. And the people murmured against Moses, saying, What shall we drink? And he cried unto

* FAURÉ, *Annales de Chem. et de Phys.*, Septembre, 1853, p. 84.

the Lord, and the Lord showed him a *tree*, which when he had cast into the waters, the waters were made sweet."*

As in our European sandy dunes, the waters of the sandy wilderness may contain an albumen-like substance which an astringent plant will coagulate. The discovery of such a plant among the natural vegetation of the desert would give, therefore, the means of purifying and rendering it wholesome, as cuttings of the oak tree render salubrious the waters of the Landes of La Gironde.

5°. Water, also, absorbs or dissolves different kinds of air or gas in different proportions; and upon this property depend some things which are familiar to us in common life, and which, therefore, it may be proper to mention. Thus—

First. It absorbs its own bulk of carbonic acid gas—and it does so under every pressure.

The meaning of this is explained as follows. We take a strong, tall, glass jar (fig. 9), graduated into five equal divisions, and provided with an air-tight piston, *p*. Into this jar we pour pure water up to the first division (1), fill up the jar quickly with carbonic acid, fit in the piston and shake the jar. The piston will then gradually sink one division (to 4)—that is, the water will dissolve or absorb its own volume of the gas, under the ordinary pressure of the atmosphere. But if, the arrangement being as before, we apply at once to the piston rod *r* a pressure equal to another atmosphere—15 lb. to the square inch—the piston will immediately sink two divisions (to 3), or the gas will be compressed to half its bulk. If the whole be now shaken, the piston will, as at first, gradually sink one division (to 2). In other words, the water will again absorb its own bulk of the gas under this increased pressure.

Fig. 9.

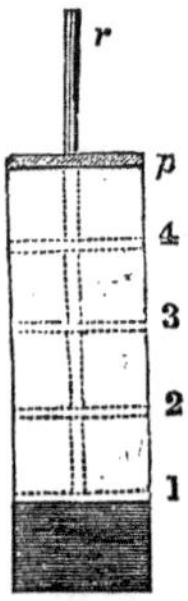

* Exodus, xv. 23.

Or, if we apply at once a pressure of three atmospheres—45 lb., making, with the ordinary atmosphere, four in all, or 60 lb. to the inch, which press upon it—the piston will sink at once three divisions (to 2), reducing the gas to one-fourth of its bulk. If, now, the water be agitated, the piston will again gradually sink one division, and the whole gas will disappear—that is, the water will again absorb its own bulk of the gas at this new pressure.

If, now, the applied pressure of 45 lb. be removed, the gas will gradually rise out of the water and force up the piston, till it finally rests, as in the first experiment, at the division No. 4, the water retaining only its own bulk of the gas at the ordinary pressure of one atmosphere.

It is because of this interesting property that, with the aid of machinery, water can be overcharged with carbonic acid in the soda-water manufactories, and that the gas escapes with so much violence from a soda-water bottle when the cork is withdrawn.

But the result is the same whether the carbonic acid be forced into the water ready prepared—as is done by the soda-water maker—or is formed in the bottle itself from substances contained in the water. The latter is the case in all fermenting liquors contained in bottles. The carbonic acid is gradually produced in the interior of the bottle during the progress of the chemical change we call fermentation. As fast as it is produced the water dissolves it, the pressure of the gas upon the inner surface of the bottle increasing at the same time. If the bottle be of sufficient strength, the only consequence is, that the cork will be forced out if not firmly tied down; or that, when the cork is withdrawn, the gas will drive out the liquor in its own eagerness to escape. If the bottle be too weak, it will be burst by the pressure, as often happens with soda-water; and, sometimes, to thousands of bottles at a time in champagne cellars. In other wines, and in beer and porter, especially when well hopped, carbonic acid is produced in smaller quantity. But it is to the presence of this gas, dissolved in this way, that the latter liquors owe their briskness when poured from the bottle, and to the natural escape of the gas that they become flat, stale, or dead, as we call it, when they are exposed to the air.

Water absorbs also the gases, oxygen and nitrogen—of which the atmosphere chiefly consists—but not in the precise proportions in which they exist in the air. We have seen that the air we breathe contains about 21 per cent. of oxygen, but in the air which we can extract from water it exists to the amount of 31 to 33 per cent. This, among other purposes, is an adaptation to the wants of fishes, and generally of those marine animals which extract the oxygen they require for the support of life, from the water in which they live. They can obtain the necessary supply of this gas more easily from air which contains one-third than from one which contains only one-fifth of this vital principle. If proof of this were required, it is found in the observation that, where circumstances have been such as to deprive river water of a portion of its oxygen, the fish have been found dead in great numbers.

It has recently been discovered by Hayes, that the water of the sea contains more oxygen near its surface than at a depth of one or two hundred feet. This is probably connected with the comparative scarcity of animal life at great depths.

This tendency of water to dissolve more oxygen, in proportion to the nitrogen, than exists in common air, explains another curious circumstance which long puzzled philosophers as well as ordinary people. If a bottle be filled quite full with snow, be well corked, and then put into a warm room, the snow will melt, and the bottle will be filled, perhaps, one-third with water and two-thirds with air. If this air be examined, it will be found to contain less oxygen than atmospheric air—sometimes not more than 12 or 14 per cent.; while atmospheric air, as we have seen, contains 21 per cent. Hence it was long supposed that the air, always present in snow, naturally contained this small proportion of oxygen, and that snow, therefore, possessed some peculiar property of absorbing the gases of the atmosphere in this new proportion. But the explanation is, that the snow, in melting into water, takes up a larger proportionate quantity of the oxygen than it does of the nitrogen of the air which was contained in its pores, and consequently leaves a smaller proportion behind.

Thus the water we drink, like the air we breathe, is a substance of much chemical interest. Both are indispensable to the existence of life; both are mixed in nature with many substances not essential to their composition; and both, in their most important properties, exhibit many direct relations to the growth of plants and to the wants and comforts of living animals.

CHAPTER III.

THE SOIL WE CULTIVATE.

General origin of soils; natural differences in their quality; how it arises.—Stratified and unstratified rocks.—Soils of the stratified rocks.—Improved soils where different rocks intermix.—Soils of the granites, traps, and lavas.—Agency of rains, winds, and vegetable accumulations in producing diversities of soil.—General chemical composition of soils.—Illustrations afforded by the Atlantic border of the United States.—Some plants affect sandy soils, others clay soils, and yet do not always flourish upon them.—Cause of this.—Minute chemical composition of the soil; its mineral and organic parts.—Chemical difference between granite and trap soils.—Dependence of fertility on chemical composition.—Influence of rain and moisture, and of the degree of warmth, on comparative fertility.—District floras and crops.—Influence of man in modifying geological, chemical, and climatic tendencies.—Progress of exhausting culture in new regions; example of North America.—Reclaiming influences of human exertion; example of Great Britain.

In immediate importance to man, the soil he cultivates is scarcely inferior to the air he breathes, or the water he drinks. Upon the plants which the soil produces he and all other animals depend for their daily sustenance. Hence, where the soil is fruitful, animal life is abundant; where it yields only sparingly, animals are few, and human inhabitants, as a general rule, but sparsely scattered.

The soil is formed, for the most part, from the rocks of which the crust of the earth is composed. By the action of air and water these rocks crumble, and their surface becomes covered with loose materials. The seeds of plants are sprinkled over them by the winds; they germinate and grow up; animals come to feed upon them; both plants and animals die; and thus a mixture of decayed rock, with the remains of animals and plants, gradually overspreads the entire surface of the dry land. It is to this mixture that we apply the name of soil.

But the soil thus naturally formed differs in quality, from various causes. The rocks which crumble differ in chemical composition; their crumbled fragments are spread over the surface, and sorted by wind and water in different ways; and the kind and quantity of the animal and vegetable matters they are mixed with differ much. Through the agency of these and similar causes of diversity, many varieties of soil are produced, which are not only unlike to each other in their sensible properties, but very different also in their agricultural value.

If we examine with a little attention the numerous rocks we meet with in travelling over a country like our own, an important difference in their physical structure will early strike us. Some are seen to form hills, cliffs, or mountains, which consist each of a single huge lump or mass, cracked here and there, perhaps irregularly, but exhibiting no continuous division into distinct parts or portions. Others again are as clearly divided into layers or beds, spread over each other like vast flagstones of different thicknesses, sometimes extending horizontally for distances of many miles. The following section (fig. 10) exhibits these differences of physical appearance.

Fig. 10.

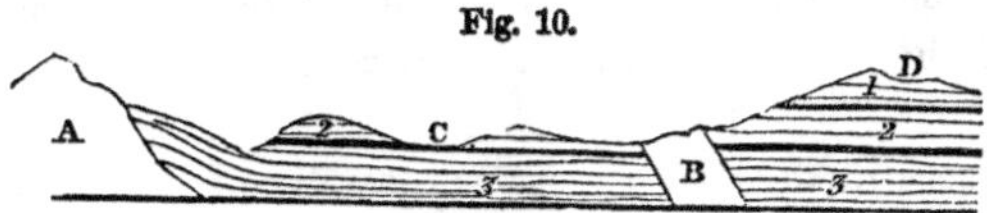

The rocks marked A and B are the undivided masses, those marked C D are the rocks which lie in beds. The numbers 1 2 3 indicate the groups into which the beds, when numerous on any spot, can usually be subdivided.

The most ignorant of science can observe differences of this kind—it requires only the use of the eyes; and yet this difference of structure is so important, that upon it is founded the division of all rocks into *stratified* and *unstratified*. Those which are composed of beds or strata are called stratified, those in which no such partings are visible are called unstratified.

The stratified rocks cover by far the largest portion of the earth's surface. They are not always quite horizontal, as represented in the above section; they are more often inclined, so as to dip into the earth at a greater or less angle. Sometimes they are even piled against each other like flagstones placed on edge. The following section (fig. 11) ex-

Fig. 11.

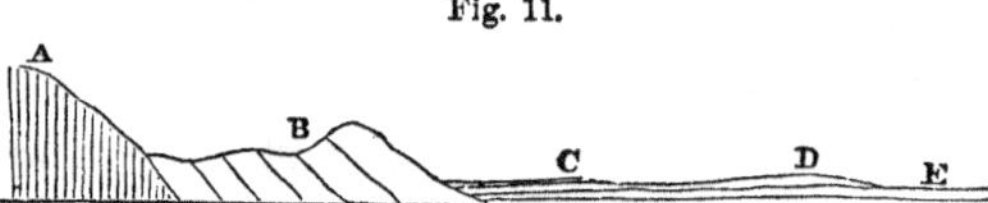

hibits these three several modes in which the stratified rocks occur, A showing them on edge, B dipping at a considerable angle, and C D E perfectly horizontal. This disposition of the rocks, it will be seen, must materially affect the quality of the soil, and especially the extent of surface over which any particular soil is to be found. If the quality of the soil depend in any degree upon the quality of the rock, the changes of soil must be very frequent where the surface is formed of the edges only of different rocks, as is seen at A and B.

These stratified rocks consist essentially of one or more of three different kinds of matter only: limestones, sandstones, and clays, more or less hard, form the substance of them all. When a limestone crumbles, it forms a calcareous soil; a sandstone, a sandy soil; and a hard clay rock, a more or less tenacious clay soil. Hence, these are the three leading qualities of soil known and spoken of among practical men.

But many rocks do not consist altogether either of limestone, of sandstone, or of clay, but of a mixture of each in varied proportions. The crumbling of such rocks, therefore, gives rise to soils of various intermediate qualities, neither calcareous, properly speaking, nor sandy nor clayey; and these form, for the most part, those more open, fertile, and valuable loams, which the farmers of every country prefer to cultivate.

Similar mixed soils are also naturally produced where the edges of different rocks overlap each other, and mingle their mutual debris. Thus, when the fragments of a rock rich in lime naturally intermix with one poor in this ingredient, the soil produced is of a much better and more useful quality than when the surface is formed by the fragments of one of the rocks only. This is illustrated in the south of England in many places, where the materials of the plastic clay, the chalk, and the green-sand, meet and intermingle, as seen in the following section, (fig. 12).

Fig. 12.

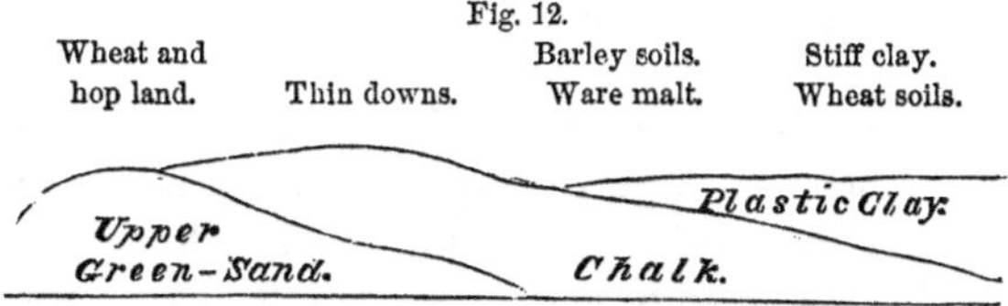

This woodcut represents the plastic clay as coming in contact with the chalk which lies below it, and the chalk again coming in contact with the upper green-sand, upon which it rests. At the first point of contact the heavy difficult clays change into open barley soils, producing a grain which, for quality and malting properties, is not excelled by any in the kingdom. And, again, at the contact of the chalk and upper green-sand, the mixed soil is equally celebrated for its crops of wheat, and for the fertility of its hop-gardens.

The unstratified rocks, again, consist chiefly of three varieties—the granites, the traps, and the lavas. These rocks also crumble more or less rapidly, and produce soils which, in granitic countries, are generally poor, over trap-rocks generally rich, and upon decayed lavas, often remarkable for fertility. In the granite districts of Devonshire and Scotland we see the poor soils which this rock produces, and in the low country of Scotland, and in the north of Ireland, the rich soils of the trap. Italy and Sicily, and every other volcanic country in the Old World, exhibit in their soils the fertilizing influence of the modern lavas.

In new countries the same phenomena reappear, similar rocks everywhere producing similar soils. Thus, at the base of the famous gold-bearing mountains of Victoria, stretches "a fertile and beautiful country—the garden of Australia Felix —the rich soil of which is the product of decomposed lava."* And for ages, probably, after the gold mines have been forgotten, these rich park-like plains will continue to yield luxuriant harvests of golden grain to the industrious cultivator.

But the earth's surface is varied with hill and valley, mountain and plain, so that the rains which fall are able to flow along the slopes, and to gather themselves into rivulets, streams, and rivers. In so flowing they wash out the finer and lighter particles from among the fragments of the crumbled rocks, and carry them into the valleys and plains.

* *Quarterly Journal of the Geological Society*, ix. 75.

The constant repetition of this washing gradually sorts the fragments of each rock, spreading the finer portions on the lower ground and along the courses of rivers, and leaving on the hills and slopes the coarser and less easily transported materials.

Hence from the same rock different varieties of soil arise. Coarse sands and gravels may overspread the higher ground, while fine sand, clays, or loams, cover the plains or valleys beneath. From a mixed stratified rock the clay or lime may be washed out and spread over the low plains, leaving only a poor and barren sand on the slopes above; or from a decaying granite the felspar-clay may be washed down, leaving the hungry and unfertile quartz to cover the naked rock.

In some countries, winds play a similar part. They lick up the fine dust as they sweep over a country, and carry it often far away to other regions; or, rushing from the sea, they bear inland the sands of the shore, and cover with sandy downs or barren deserts soils which are naturally rich and productive in vegetable food.

Thus physical causes modify the quality of the soils which different rocks naturally tend to produce. They assort or re-arrange the materials of which a rock consists, and they often bear to great distances, and spread over other rocks, the finer particles into which it crumbles. The so-called alluvial soils, which border so many of our rivers, are produced by such a sorting, produced through the agency of water. The sandy downs of European countries, and many of the desert regions of Africa and Asia, owe their existence to the sorting agency of the wind.

Vegetation also has its influence. When a tree or humbler plant dies on a dry surface, it gradually decays, and disappears into the air. Let it be immersed in stagnant water, and it blackens, falls to pieces, and crumbles, perhaps, but in substance long remains where it fell. Let others grow up, die, and fall on the same moist spot, and the black vegetable matter will accumulate from year to year. In this way, where shallow water rests on an impervious bottom, peat-bogs and other collections of vegetable matter gradually cover the surface. They bury the fragments of the crumbled rocks sometimes under a great depth of vegetable matter, and form those unmanageable peaty soils which overspread so large a portion of Scotland, and especially of the north and west of Ireland.

Such are the principal natural causes of diversity in soils. In the chemical composition of the rocks we recognise the fundamental or primary cause; in the physical distribution of rains and winds, and in their mechanical action, an important secondary cause; and in the growth and accumulation of vegetable matter, a third more special and less widely operating agent in the production of such diversities.

By these agencies are formed the varieties of soil generally described as sandy soils, clay soils, limestone or marly soils, and peaty soils. These terms all indicate important chemical differences, though practical men have hitherto had their attention too little drawn to the influence which chemical composition exercises over agricultural value. The sandy soil is distinguished by consisting chiefly of quartzose or silicious sand—another form of flint, rock-crystal, or the substance which chemists call *silica;* the limestone or marly soil, by containing much limestone, chalk, or other variety of what chemists distinguish as *carbonate of lime;* the clay soils, by abounding in clay, a compound substance, consisting chiefly, besides silica, of a substance to which chemists give the name of *alumina.*

But the economical value of a soil is often naturally affected by physico-geological considerations, which are altogether independent of the chemical composition of the rock from which it is formed. The mere physical character of the rock, for example, from which the soil is formed, often determines not only the kind of husbandry which can be profitably followed, but the class of farmers by whom the land is to be occupied, and even whether it can be profitably cultivated at all. The chalk rocks present an illustration of this. These are in most countries very porous and absorbent. Wells sunk into them yield no water, and superficial pits, to receive and retain the rain water, are the main resource of the inhabitants. This, with the thin soils and short grass of our chalk downs, has long determined the conversion of the chalk wolds into extensive sheep-walks. But in countries, by climate and otherwise, unsuited to sheep, and where the little rain that falls is soon licked up by the heats of summer, this use of the land becomes impossible, and an artificial supply of water becomes indispensable to the existence of permanent and extended cultivation. To obtain this, deep wells sunk through the chalk are the only available resource, and this at once determines that the possessors must be men of large means, or at least that the land must be worked by a class of wealthy cultivators. The upper portion of the State of Alabama, in North America, is in this condition. Situated on the porous chalk, it is destitute of surface water, unless where the rivers pass. In a hot climate, its herbage is burned up in summer, so that it is unsuited for a pastoral husbandry. It grows some flinty wheat, but it is almost equally unsuited to be an extensive producer of grain. Devoted chiefly to the cotton culture, it is held in large properties, and hundreds of deep Artesian wells already riddle the country, and yield the needful supplies of water.

The following section (fig. 13) of the Atlantic coast-line of North America, from the sea to the mountains, will serve

Fig. 13.

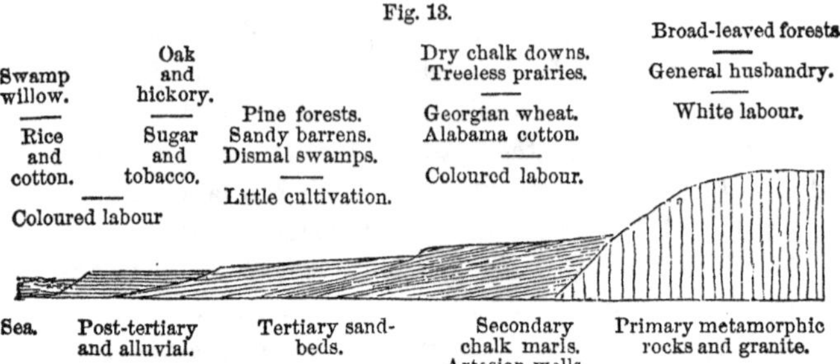

to illustrate nearly all the points I have brought under the notice of the reader in the preceding part of this chapter, in reference at least to the stratified rocks. This section shows :—

1°. How, over large tracts of country, the rocks are seen to be at different angles of inclination; some, as in the high land to the right, standing on their edges; and some, as the layers of alluvial soil on the sea-shore, lying nearly on a level.

2°. How, over extended areas, the surface rock may consist chiefly of clay, as in the post-tertiary and alluvial deposits near the sea; of sand, as in the tertiary beds; of limestone, as in the chalk marls; and of mixed materials, as on the hills, where numerous thin beds resting on their edges rapidly succeed each other.

3°. How the character of the soil changes distinctly with the surface rock—being rich and productive on the post-tertiaries, sandy and barren on the tertiaries, dry and chalky on the secondary marls, useful and loamy on the slopes of the older mixed and metamorphic rocks.

4°. How the natural vegetation and the artificial pro-

* The word metamorphic here used means changed or altered—as clay, for example, is changed when it is baked into tiles or bricks.

duce of the soil vary in like manner; and how the kind of husbandry, and we might almost say the social state, is determined by the character of the dead rocks. It is certain, at least, that the *profitable employment* of slave instead of free labour depends very much upon the character of the superficial rocks, of the soils they yield, and of the crops they can readily be made to grow.

5°. And lastly, how dismal peaty swamps disguise the natural character of the surface in some regions; and how the want of water in others renders profitable cultivation impossible, unless, by expensive borings, it can be brought up from great depths.

The amount of chemical knowledge embodied in the general chemical description of soils already given, is useful and satisfactory as explaining their general origin, and is sufficient even to direct the practical man in reference to certain economical operations. Long experience and observation, for example, have made generally known to practical men that certain cultivated plants and trees prefer to grow best upon sandy soils, others on limestone soils, others on clay soils, and others again on soils of a mixed or loamy character. If one of these trees or plants is to be grown, therefore, a sandy or other soil suited to it is sought for; or if a sandy or clay soil is to be profitably planted or cultivated, the tree is selected which has been seen to flourish, or the crop which has yielded profitable harvests on other sands or clays of a similar kind.

But when we come to inquire more particularly into the relations between plants and soils, this elementary chemical knowledge fails us. The same plants do not flourish on all sands, on all clays, or on all marls equally. Why is this? Or the trees flourish for a while, and then die out; or the crop for a few years yields remunerative returns, and then ceases to give a profitable harvest. How are these changes to be explained? The soil is as sandy, the clay as stiff, and the marl as rich in lime as ever, and yet the plants which formerly rejoiced in the several soils now refuse to grow in them!

A more minute chemical examination answers these questions, and in each case suggests a remedy for the evil complained of. This examination shows—

First, That when a weighed portion of perfectly dried soil, of any kind on which plants are capable of growing, is heated to redness in the air, a part of it burns away, and what is left is found to have sensibly diminished in weight. The combustible portion which thus disappears consists of the animal and vegetable (or organic) matter, of which all soils contain a sensible quantity. In some the proportion is very small, as in the sandy soil on which the cinnamon tree grows at Colombo, in Ceylon, which contains only one per cent. of organic matter. In others it is very large, as in our own peaty soils, many of which lose upwards of three-fourths of their weight when burned in the air.

Second, That the earthy incombustible part of the soil—besides the silica of the sandy soils, the alumina of the clays, and the lime of the marly soils—contains various other substances, occasionally in large proportions. Among these, potash, soda, magnesia, oxide of iron, sulphuric acid, and phosphoric acid,* are the most important.

In all soils upon which plants grow well and in a healthy manner, every one of these substances exists. If they are altogether absent, the plant refuses to grow. If they are present in too small quantity, the plant will be stunted and unhealthy. If the same kind of plant be grown for too long a time in the same soil, one or more of these substances will

* Sulphuric acid, so called from its containing sulphur, is the name given by chemists to oil of vitriol; and phosphoric acid is the white substance produced when phosphorus is burned in the air.

become scarce, either absolutely, or in a form in which the plant can take them up; and hence the roots will be unable to obtain as much of them as the health and growth of the plant requires. It is plain enough, therefore, why plants often refuse to grow even on the kind of soils they especially prefer, and why, having grown well on them for a while they refuse to do so any longer. The soil does not contain all they require for their support, and in the proper form or having once contained them all in sufficient proportions, it does so no longer. And the remedy for this special evil is equally clear. Add to the soil the mineral ingredients which are deficient, or introduce them in an available form, and the plant will spring up with its old luxuriance.

In like manner, that part of the soil which burns away—the organic part—when minutely examined, is found to consist of numerous different forms of matter. These are all included, however, in one or other of two groups—those which contain the element nitrogen, described in the first chapter,* and those which contain none of this element. All soils in which plants grow well, must contain and be able to yield to the plant a sufficiency of the substances belonging to each of these groups, and especially of those which contain nitrogen. If they do this too sparingly, the plant will become sickly; if they withhold them altogether, the plant will die.

It is with the organic as with the mineral constituents of the soil, therefore: they may be present too sparingly, and thus the sand-loving plant may refuse to grow even in a sandy soil, or one which loves lime where lime abounds. It may refuse to grow even when all the mineral matters it requires are abundant in the soil, because the necessary organic food is still wanting. The full remedy, therefore, is obtained only when we supply to the unproductive soil the

* See THE AIR WE BREATHE.

necessary organic as well as the necessary inorganic or mineral matters of which it may stand in need.

I may in some measure illustrate this by referring to a special case, common in nature, and to which I have already alluded in the present chapter. The granitic rocks, I have said, produce generally poor, the trap rocks, on the other hand, generally fertile soils. To what difference in the mineral matter of the rocks is this economical difference in the soils chiefly to be ascribed ?

If a piece of each of the two kinds of rock be submitted to analysis, a remarkable but almost constant difference is discovered in their comparative composition. Besides the silica and alumina of which I have already spoken as existing in clays, the granites contain a copious supply of potash and soda, with occasionally minute quantities of magnesia, lime, and oxide of iron. The traps, on the other hand, abound in all these ingredients nearly equally; and as experience has shown that the presence of all, in sensible proportion, is necessary to make a soil fertile, the reason of the natural difference between granite and trap soils becomes at once apparent. The one is defective, while the other abounds in the mineral constituents of a fertile soil. And the means for improving the granite soils become equally apparent. Add, as a first step, the mineral substances in which granite is deficient, and fertility may gradually ensue. It is for this reason that in granite countries the application of lime, in some of its forms, is a favourite practice—one discovered to be remunerative long before chemistry had shown the reason why.

Although, therefore, the first use of the soil in reference to the general vegetation of the globe is to afford to plants a firm anchorage, so to speak, for their roots—and although the growth of many useful plants seems at first sight to be dependent on the rude and general question only, as to whether the soil they occupy be a sand, a clay, or a calcareous marl,—yet a minute chemical examination shows that their usefulness to plants is in reality dependent upon the presence of a large number of chemical substances, both of mineral and of organic origin. If these are present, any plants will grow upon them that are suited to their mechanical texture and to the climate of the place. If they are absent, whatever be the texture of the soil, and whatever the climate, the plant will languish and die. And the whole art of manuring consists in adding to the soil those things in which it is deficient—at the right time, in a proper chemical condition, and in the requisite proportions. What services, chemical and physiological, the several constituents of the fertile soil really render to the plant that grows upon it, will appear in the succeeding chapter.

But suppose all the necessary chemical adjustments to be made—the composition of the soil, that is, to be such as is usually attendant upon fertility—physical conditions and agencies often intervene to falsify the predictions of chemistry. Thus, the fall of rain may be too small to keep the land in that condition of moisture which is required for the growth of plants. Hence the wide and naked deserts which extend over the rainless regions of the earth's surface. Whatever be the chemical composition of the soil in these regions, vegetation is impossible, and the labour of man, except he bring in water, almost in vain. Or the surface of a country may be so flat that the rains which descend upon it can find no outlet. They stagnate, therefore, and render it unpropitious to the cultivator, so that fertility cannot show itself, whatever the soil may contain, unless an easy escape for the superfluous water be first provided. Or the rains may fall unseasonably, as they do in Iceland, where they appear in the autumn, when the barley should be ripening, in far too copious showers to permit even this hardiest of grain crops to be cultivated with profit in the island.

So the thermal conditions of a region may interfere with its fertility. Abstract chemistry says, "Let the soil contain the necessary constituents, and any crop will grow upon it." But physiology modifies this broad statement, by showing, *first*, that whatever be the chemical composition of the soil, it must possess a certain physical texture before this or that plant will grow well upon it. That which naturally affects a clay soil will not grow well upon a sand; so one which delights in a blowing sand will languish in a moorish peat, however rich in chemical ingredients it may be. And, *second*, that the temperature or warmth of a place determines equally whether its naturally rich soils shall grow this crop or that. Upon the combined influences, in fact, of moisture and warmth, which make up what we call climate, depend in a great degree the varied floras and cultivated crops of the different regions of the globe. Thousands of plants, which beneath the tropics produce abundantly, will in the same soil scarcely expand a flower when placed beneath an arctic sky.

However important, therefore, the geological origin of a soil and its chemical composition may be, where climate is favourable neither are able to effect anything in the way of raising food for man, where a duly attempered moisture and warmth are wanting.

But man also exercises an influence on the soil, which is worthy of attentive study. He lands in a new country, and fertility everywhere surrounds him. The herbage waves thick and high, and the massive trees raise their proud stems loftily towards the sky. He clears a farm from the wilderness, and ample returns of corn pay him yearly for his simple labours. He ploughs, he sows, he reaps, and from her seemingly exhaustless bosom the earth gives back abundant harvests. But at length a change appears, creeping slowly over and gradually dimming the smiling landscape. The corn is first less beautiful, then less abundant, and at last it appears to die altogether beneath the resistless scourge of an unknown insect, or a parasitic fungus.* He forsakes, therefore, his long cultivated farm, and hews out another from the native forest. But the same early plenty is followed by the same vexatious disasters. His neighbours partake of the same experience. They advance like a devouring tide against the verdant woods. They trample them beneath their advancing culture. The axe levels its yearly prey, and generation after generation proceeds in the same direction —a wall of green forests on the horizon before them, a half desert and naked region behind.

Such is the history of colonial culture in our own epoch; such is the vegetable history of the march of European cultivation over the entire continent of America. From the shores of the Atlantic, the unrifled soil retreated first to the Alleghanies and the shores of the great lakes. These are now overpast, and the reckless plunderer, axe in hand, scarcely retarded by the rich banks of the Mississippi and its tributary waters, is hewing his way forward to the Rocky Mountains and the eastern slopes of the Andes. No matter what the geological origin of the soil may be, or what its chemical composition; no matter how warmth and moisture may favour it, or what the staple crop it has patiently yielded from year to year, the same inevitable fate overtakes it. The influence of long-continued human action overcomes the tendencies of all natural causes.

I need scarcely refer, as special examples of this fact, to

* In New England and the British provinces of North America the wheat is overwhelmed by the *fly*; in New Jersey and Maryland, the wide peach-orchards by the *borer*, and a mysterious disease called the *yellows*; and in Alabama the cotton plant by the *rust*.

the tracts of abandoned land which are still to be seen along the Atlantic borders of Virginia and the Carolinas. It is more interesting to us to look at those parts of America which lie farther towards the north, and which, in modes of culture and kinds of produce, more nearly resemble our own.

The flat lands which skirt the lower St. Lawrence, and which near Montreal stretch into wide plains, were celebrated as the granary of America in the times of the French dominion. Fertile in wheat, they yielded for many years a large surplus for exportation; now they grow less of this grain than is required for the consumption of their own population. The oat and the potato have taken the place of wheat as the staples of Lower Canadian culture, and as the daily sustenance of those who live on the produce of their own farms.

So, in New England, cultivation of wheat has gradually become unprofitable. The tiller of the worn-out soils of this part of the United States cannot compete with the cultivator of the fresh land yearly won by the axe and the plough from the western wilderness, and he is fain to betake himself to the raising of other crops. The peculiarly wheat-producing zone is yearly shifting itself more completely towards the west. This has long been evident to the careful observer, and to the collector of statistical data. I brought it distinctly before the public in my work on North America.* And a striking proof of the correctness of my views is afforded by the subsequent returns of the United States census of 1850. From these it appears that, while the produce of wheat in the New England States in 1840 amounted to 2,014,000 bushels, it was reduced in 1850 to 1,078,000 bushels. So rapidly, even now, is the influence of human agency on the natural tendencies of the soil, continuing in these countries to manifest itself.

* *Notes on North America*, vol. i. chap. xiii.

But the influence of man upon the productions of the soil is exhibited also in other and more satisfactory results. The improver takes the place of the exhauster, and follows his footsteps on these same altered lands. Over the sandy, forsaken tracts of Virginia and the Carolinas he spreads large applications of shelly marl, and herbage soon covers it again, and profitable crops. Or he strews on it thinner sowings of gypsum, and as if by magic the yield of previous years is doubled or quadrupled.* Or he gathers the droppings of his cattle and the fermented produce of his barnyard, and lays it upon his fields—when, lo! the wheat comes up luxuriantly again, and the midge, and the rust, and the yellows, all disappear from his wheat, his cotton, and his peach trees!

But the renovator marches much slower than the exhauster. His materials are collected at the expense of both time and money, and barrenness ensues from the easy labours of the one far more rapidly than green herbage can be made to cover it again by the most skilful, zealous, and assiduous labours of the other. But nevertheless, among energetic nations, this second tide follows inevitably upon the first, as they advance in age, in wealth, and in civilisation. Though long mismanagement has, in a minor sense, desolated large portions of north-eastern America, a new fringe of verdant fields has already begun to follow towards the west, though at a long interval, the fast-retiring green belt of the virgin forests. A race of new cultivators, taught to treat the soil more skilfully, to give their due weight to its geological origin, to its chemical history, to the conditions of climate by which it is affected, and to the reckless usage to which it has so long been subjected—this new race

* For examples of both these results, see the *Essay on Calcareous Manures*, by Edward Ruffin, the publication of which in Virginia, in 1832, marks an epoch in the agricultural history of the slave states of North America.

may—*will*, I hope, in time—bring back the whole region to more than its original productiveness. Both the inherited energy of the whole people, and the efforts which State agricultural societies, and numerous zealous and patriotic individuals in each State are now making, justify us in believing that such a race of instructed men will gradually spread itself over the rural districts in every part of the Union. The previous success of the mother country guarantees a similar successful result to their kindred exertions.

For we have not to go far back in the agricultural history of Great Britain to find a state of things not much different from the present condition of the land in North America. We require to turn aside but a short way from the high-road, in some districts of England, still to find in living operation nearly all the defects and vices of the present American system of farming.* A century and a half has, I may say, changed the whole surface of our island. But what labour has been expended, what wealth buried in the soil, what thought lavished in devising means for its recovery from long-inflicted sterility! Commerce has brought in from all parts of the world new chemical riches, to replace those which a hundred previous generations had permitted rains and rivers to wash out of the soil, or to carry away to the sea. Mechanical skill has given us the means of tilling the surface economically, of bringing up virgin soils from beneath, and of laying dry that which over-abundant water had prevented our forefathers from utterly impoverishing; and scientific investigation has taught us how best to apply all these new means to the attainment of the desired end.

It may be said, with truth, that Great Britain at this moment presents a striking illustration of the influence of man in increasing the productiveness of the soil. This ex-

* See, for instance, the state of farming in Lancashire, as described in the *Royal Agricultural Journal*, vol. x. part I.

ample guarantees, as I have said, the success of similar operations in the United States of America and in our British colonies; while the now advanced condition, especially of our chemical knowledge, both in regard to the soil which is to be cultivated and to the plants we wish to grow, insures a far more easy and certain advance to the process of restoration in these countries than in past times could take place among ourselves; less waste of time and money in ill-adjudged experiments, and less cost of labour in all the necessary operations of husbandry.

CHAPTER IV.

THE PLANT WE REAR.

A perfect plant, what.—Effects of heat upon it.—Contains carbon, water, and mineral matter.—Relations of the plant to the air.—Structure of the leaf.—Its pores absorb carbonic acid, and give off oxygen gas.—Relations to water.—Structure of the root.—Purposes served by water.—Relations to the soil.—Plants affect peaty, sandy, loamy, or clay soils.—Effects of the drain, of lime, or of manure.—The art of manuring.—How the colours of flowers may be changed.—Effect of culture upon wild plants.—The carrot, the cabbage, the turnip.—Garden fruits, flowers, and vegetables.—Origin of wheat and its varieties.—How these changes are produced.—Plants which follow the footsteps of man; why they follow him.—Rapidity of growth in favourable circumstances.—The yeast plant in grape juice.—Manufacture of dry yeast.—Chemical changes within the plant.—Production of numerous peculiar substances—medicines, perfumes, and things useful in the arts.—The green of the leaf, and the poison of the nettle.—The covering of the ripe potato, apple, and young twig.—General purposes served by vegetation.—It adorns the landscape.—In relation to dead nature, it purifies the atmosphere, produces vegetable mould, and forms deposits of combustible matter.—In relation to living animals, it supplies subsidiary luxuries and comforts, but its main use is to feed them.—Numerous interesting chemical inquiries suggested by the natural diversities and different effects of the vegetable food consumed by herbivorous and omnivorous races.

A FAMILIARITY with the chemical relations of the plant we rear makes still more apparent the relations of chemistry to the soil we cultivate.

A perfect plant consists essentially of two parts—the stem and the leaf. The root is an underground extension of the stem, as the bark is a downward prolongation of the leaf. The several parts of the flower, also, are only changed leaves.

When any part of a plant is heated in a close vessel, it

gives off water, vinegar, and tarry matters, and leaves behind a black, bulky, coaly mass, known by the name of wood charcoal; or if billets of wood be heaped up in the open air, covered carefully over with sods, and *smother*-burned, as it is called, with little access of air, the tar and other matters escape into the atmosphere, while the charcoal remains undissipated beneath the sod. This charcoal is an impure form of carbon. The manufacturer of wood-vinegar collects the volatile substances as the more important products. The charcoal-burner allows them to escape, the black residue being the object of his process. Both experiments, however, are the same in substance, and both prove that carbon and water form large parts of the weight of all plants.

If a piece of wood charcoal be burned in the air it gradually disappears; but when all combustion has ceased, there remains behind a small proportion of ash. The same is seen if a portion taken from any part of a living plant be burned in the air. Even a bit of straw kindled in the flame of a candle (fig. 14), and allowed to burn, will leave a sensible quantity of ash behind. All plants therefore, and all parts of plants, besides water and carbon, contain also a sensible proportion of mineral inorganic matter which is incombustible, and which remains unconsumed when they are burned in the air.

Fig. 14.

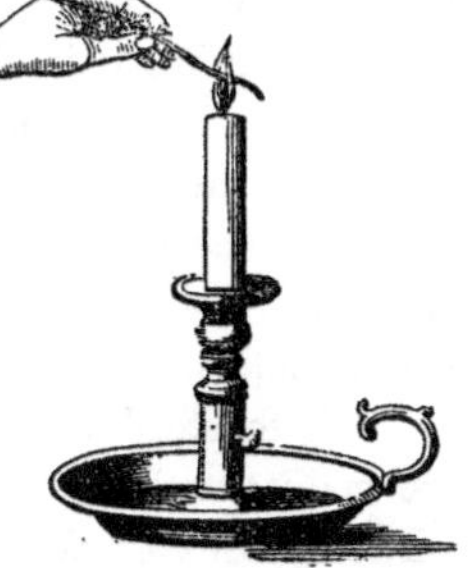

The carbon of the plant is chiefly derived from the air, the water and the mineral matter chiefly from the soil in which it grows. Thus the plant we rear has close chemical relations with the air we breathe, with the water we drink, and with the soil we cultivate. I shall briefly illustrate these several relations in their order.

First. The plant is in contact with the air, through its leaves and its bark. The surface of the leaf is studded over with numerous minute pores or mouths (*stomata*), through which gases and watery vapour are continually entering or escaping, so long as the plant lives. In the daytime they give off oxygen and absorb carbonic acid gas. During the night this process is reversed—they then absorb oxygen and give off carbonic acid.

We have already seen that carbonic acid consists of carbon and oxygen.* It is from the large excess of this gas which plants absorb during the day that the greater part of the carbon they contain is usually derived.

The number and activity of the little mouths which stud the leaf are very wonderful. On a single square inch of the leaf of the common lilac as many as 120,000 have been counted; and the rapidity with which they act is so great, that a thin current of air passing over the leaves of an actively-growing plant is almost immediately deprived by them of the carbonic acid it contains.

The gas thus absorbed enters into the circulation of the plant, and there undergoes a series of chemical changes which it is very difficult to follow. The result, however, we know to be, that its carbon is converted into starch, woody fibre, &c., to build up the plant, while its oxygen is given off to maintain the purity of the air.

These pores of the leaf absorb also other gaseous substances in smaller quantity—such as ammonia, when it happens to approach them; and especially they absorb watery vapour, when previous heat or drought has dried the plant, and made the leaves droop soft and flaccid. Hence the natural rain enlivens and invigorates the herbage, and the artificial shower gives new life to the tenants of the conservatory. The falling water not only supplies their want of

* See The Air we breathe.

fluid, but it washes also the dusty surface of the leaves, and clears their many mouths, so that with fresh vigour they can suck in new nourishment from the surrounding air.

The green bark of the young twig is perforated with pores like the green leaf, and acts upon the air in a similar way; but as it hardens and gets old the pores become obliterated, and it ceases to aid the leaves in absorbing carbonic acid, or in giving off oxygen to the atmosphere.

Second. The water which fills the vessels of the plant, though partly derived from the air in seasons of drought, and drunk in by the leaves from the dews and falling showers, is principally sucked up by the roots from the earth in which it grows. These roots, as I have said, are only downward expansions of the stem. At the surface of the ground they exhibit a bark without and a pith within the woody portion. But as they descend, these several parts disappear, and graduate into a porous, uniform, spongy mass, which forms the ends of the fibry rootlets. Upon the surface of these rootlets the microscope enables us to perceive numerous minute hairs which, like hollow horns, thrust themselves laterally among the particles of the soil. Through these hollow hairs, as it is believed, the plant draws from the earth the supplies of water it constantly requires, and which in droughty weather it so copiously pours out from its leaves into the air.

How interesting it is to reflect on the minuteness of the organs by which the largest plants are fed and sustained. Microscopic apertures in the leaf suck in gaseous food from the air; the extremities of microscopic hairs suck a liquid food from the soil. We are accustomed to admire, with natural and just astonishment, how huge rocky reefs, hundreds of miles in length, can be built up by the conjoined labours of myriads of minute insects labouring together on the surface of a coral rock; but it is not less wonderful that, by the ceaseless working of similar microscopic agencies in leaf and root, the substance of vast forests should be built up, and made to grow before our eyes. It is more wonderful, in fact; for whereas in the one case dead matter extracted from the sea is transformed only into a dead rock, in the other the lifeless matters of the earth and air are converted by these minute plant-builders into living forms, lifting their heads aloft to the sky, waving with every wind that blows, and beautifying whole continents with the varying verdure of their ever-changing leaves.

The water which the roots absorb, after it has entered the plant, serves many important physiological and chemical purposes. It fills up mechanically and distends the numerous vessels; it mechanically dissolves, and carries with it, as it ascends and descends, the various substances which are contained in the sap; it moistens and gives flexibility to all the parts of the plant, and, by evaporation from the leaves, keeps it comparatively cool, even in the sunniest weather. But its chemical agencies, though less immediately sensible, are equally important. It combines with the carbon, which the leaf brings in from the air, and forms woody fibre, starch, and gum—all of which consist of carbon and water only; it serves as a constant and ready storehouse, also, for the supply of oxygen and hydrogen which are required, now here and now there, for the formation of the numerous different substances which, in smaller quantity than starch or woody fibre, are met with in the different parts of the plant. Thousands of chemical changes are every instant going on within the substance of a large and quickly-growing tree, and in nearly all these the constituent elements of water—its oxygen and hydrogen—play a constant part. The explanation of these, though yet very imperfectly studied, fills up already a large division of our modern treatises on organic chemistry.

Third. To the soil the plant is perceived, even by the least instructed, to have the closest relations. To the most instructed these relations every day appear more interesting and wonderful.

I have already adverted, in the preceding chapter, to what may be called the physiological habits of plants, which incline them to grow upon soils which are more or less wet, more or less sandy and porous, and more or less heavy in the agricultural sense. Owing to these habits, every variety of soil, in every climate, supports its own vegetable tribes. Thus, of the five thousand flowering plants of central Europe, only three hundred grow on peaty soils, and these are chiefly rushes and sedges. In the native forests of northern Europe and America, the unlettered explorer hails the gleam of the broad-leafed trees glittering in the sun, amid the ocean of solemn pines, as a symptom of good land on which he may profitably settle. And so the rudest peasant at home knows that wheat and beans affect clay soils,—the humblest north German, that rye alone and the potato are suited to his blowing sands,—and the Chinese peasant, that warm sloping banks of light land are fittest for his tea plant, and stiff, wet, impervious clays for his rice. Even the slave of Alabama is aware that dry open alluvials, and porous uplands, suit best the cotton he is forced to cultivate; and the still more degraded slave of Pernambuco, that the cocoa grows only on the sandy soils of the coast—just as in his native West Africa the oil-palms flourish on the moist sea-sands that skirt the shore, and the mangroves, where muddy shallows are daily deserted by the retiring tide.

But these relations of plants become more conspicuous when we examine somewhat closely the influence of artificial changes in the soil upon the kind, the growth, and the character or appearance of the plants which spring up or are sown upon it.

Thus, when a peaty soil is drained, the heaths disappear,

and a soft woolly grass (*Holcus lanatus*) overspreads its surface. A wet clay is laid dry, and the rushes and water-loving plants are succeeded by sweet and nutritious herbage. Lime is applied, and sorrel and sour grasses are banished from the old pasture; and corn then ripens and fills the ear where formerly it languished and yielded scanty returns of unhealthy grain. Crushed bones are strewed over a meadow, and abundant milk and cheese show how the eatage of cattle has been improved—or they are drilled into the ploughed land, and luxuriant root-crops exhibit their ameliorating effect. Or guano, or the droppings of cattle, or the liquid of the farmyard, or nitrate of soda, are spread upon the scanty pasture, and straightway the humble daisy and the worthless moss—symbols of poverty—disappear, and rejoicing crops of most fragrant hay prove the close connection of the plant with the soil on which it grows.

The plant derives, as I have elsewhere said, the whole of its mineral matter from the soil, and an important portion also of that which forms its combustible part. A naturally fertile soil contains all these things in sufficient abundance, and can readily supply them to the craving roots. The waters which moisten the soil dissolve them, and the minute hairs I have spoken of suck them up, and send them through the roots and stem to the several parts of the plant. The art of manuring merely supplies to the soil those necessary forms of vegetable food in which it is deficient; and the effects which follow from the addition of manures show how closely the welfare of the plant is connected with the chemical composition of the soil. The raw materials also, which it takes up by the root, like those which enter by the leaf, undergo within the plant numerous successive chemical changes, by which they are converted into the substance of the plant itself, and are fitted for those after purposes, in reference to animal life, which, in the economy of nature, the plant fulfils.

Among the pleasing proofs of such chemical changes taking place within the plant, I may mention the effects upon the colour of their flowers, which follow from the application of certain substances to the roots of plants. Charcoal powder darkens and enriches the flowers of the dahlia, the rose, the petunia, &c.; carbonate of soda reddens ornamental hyacinths, and super-phosphate of soda alters in various ways the hue or bloom of other cultivated plants. As the dyer prepares the chemical ingredients of the baths into which his stuffs are to be dipped, and varies the one with the colour he is to give to the other—so within the plant the substances applied to the root are chemically prepared and mixed, so as to produce the new colour imparted by their means to the petals of the flower.

But such effects of chemical art are far inferior both in interest and importance to those which protracted nursing have produced upon our commonly cultivated plants. The large and juicy Altringham carrot is only the woody spindly root of the wild carrot (*Daucus carota*) luxuriously fed. Our cabbages, cauliflowers, Kohl-rabis, and turnips, in all their varieties, spring from one or more species of Brassica, which in their natural state have poor woody bitter stems and leaves, and useless spindle-shaped roots. Our cultivated potato, with all its varieties, springs from the tiny and bitter root of the wild potato, which has its native home on the sea-shores of Chili; and our apples, plums, grapes, and other prized fruits, from well-known wild and little-esteemed progenitors. Our gardens are full of such vegetable transformations.

It is so also with our corn plants. On the French and Italian shores of the Mediterranean grows a wild neglected grass known by the name of Aegilops. Transplanted to the garden or to the field, and differently fed, its seed enlarges, and, after a few years' cultivation, changes into perfect and productive wheat. From other plants originally wild like this, though as yet unknown, have come our oats and barley, and rye and maize, in all their varieties, as well as the numerous forms of the Eastern durrha, rice and millet, and of the less known quinoa of Upper Chili and Peru. It is the new chemical conditions in which the plants are placed, which cause the more abundant introduction of certain forms of food into their circulation, and the more full development, in consequence, either of the whole plant, or of some of its more useful parts.

It is with unconscious reference to these improved conditions that certain wild and useless plants attach themselves to and appear affectionately to linger in the footsteps of man. They follow him in his migrations from place to place—advance with him, like the creeping and sow thistles, as he hews his way through primeval forests—reappear constantly on his manure-heaps—spring up, like the common dock, about his stables and barns—occupy, like the common plaintain, the roadsides and ditches he makes—or linger, like the nettle, over the unseen ruins of his dwelling, to mark where his abode has formerly been. Thus, with the European settler, European weeds in hundreds have spread over all Northern America,* and are already recognised as familiar things, speaking to them of a far-off home, by the emigrants now landing in thousands on the shores of Australia and New Zealand. We cannot say that all these have followed the European. Many of them have only accompanied him, and, like himself, taken root in what has proved a favourable soil. But those which cling closest to his footsteps, which go only where he goes—which, like his cat or his dog, are, in a sense domesticated—these attend upon him, because near his dwelling the appropriate chemical food is found, which best ministers to the wants of their growing parts.

* See the author's *Notes on North America*, vol. i. p. 10

How singularly dependent the plant is upon the chemical nature of the medium in which it is placed, is beautifully illustrated by the manner in which the humblest forms of vegetation are seen to grow and propagate. The yeast with which we raise our bread is a minute plant belonging to the division of the Confervæ. If we make a thick syrup of cane-sugar, and strew a few particles of this yeast upon it, they will begin to grow and propagate, will cause minute bubbles of gas to rise, and the whole syrup gradually to ferment. But if, instead of a syrup of sugar, we take a thick solution of gum, the yeast will produce no sensible effect; it will neither propagate nor cause a fermentation. In the one case the minute plant has met with a somewhat congenial food; in the other it has found nothing on which it can live and grow.

But in the juice of ripe grapes it has a more favourable medium still. "If we filter this juice, we obtain a clear transparent liquid. Within half an hour this liquid begins to grow, first cloudy, and afterwards thick, to give off bubbles of gas, or to ferment, and in three hours a greyish-yellow layer of yeast has already collected on its surface. In the heat of the fermentation the plants are produced by millions —a single cubic inch of such yeast, free from adhering water, containing eleven hundred and fifty-two millions of the minute organisms." The annexed woodcut (fig. 15) shows the appearance of the yeast plant, as seen under the microscope when the propagation is in full activity, as sketched by Turpin. The cells or globules vary in size from $\frac{1}{75000}$ to $\frac{1}{25000}$ of an English inch.

Fig. 15.

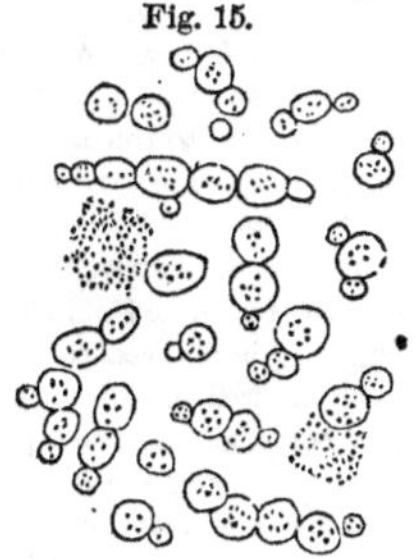

Yeast in wort for eight hours, showing—
The transparency of the yeast cells.
The granules or nuclei in their interior.
How they germinate and multiply.
How they unite into jointed filaments

The juice of the grape

thus readily propagates the seeds of yeast which accidentally reach, or are naturally present in it, because it contains the food which, in kind, in form, and in quantity, is best suited to its rapid growth.*

And so it is with larger plants in the soil. They grow well and healthily, if it contain the food in which they delight. They droop if such food is absent, and again burst into joyful life when we supply by art those necessary ingredients in which the soil is deficient.

But the special chemical changes that go on within the plant, could we follow them, would appear not less wonderful than the rapid production of entire microscopic vegetables from the raw food contained in the juice of the grape. It is as yet altogether incomprehensible, even to the most

* Whence comes the seeds of this yeast plant, which propagates itself with such wonderful rapidity? Do they exist already in the juice of the living grape? Do they cling to the exterior of the fruit, and only become mixed with the juice when it is in the wine-press, or do they float perpetually in the air, ready to germinate and multiply wherever they obtain a favourable opportunity? Whichever way they come, it would be too slow a process to wait for the natural appearance of these plants in the worts of the brewer and distiller. In these manufactories, therefore, it is customary to add a little yeast to the liquor as soon as it is considered ready for the fermentation. Then, as in the case of the grape, the growth and propagation of the plant proceed with astonishing rapidity, and large quantities of yeast are produced. This yeast in many distilleries forms an important *by-product* of the manufactory, and is collected and sold under the name of dry yeast, for the use of the private brewer and the baker. When this is done, the process adopted is nearly as follows: Crushed rye is mashed with the proper quantity of barley malt, and the wort, when made, cooled to the proper temperature. For every hundred pounds of the crushed grain, there are now added half a pound of carbonate of soda, and six ounces of oil of vitriol (sulphuric acid) diluted with much water, and the wort is then brought into fermentation by the addition of yeast. From the strongly-fermenting liquid the yeast is skimmed off, and strained through a hair sieve into cold water, through which it is allowed to settle. It is afterwards washed with one or two waters, and finally pressed in cloth bags till it has the consistence of dough. It has a pleasant fruity smell, and in a cool place may be kept for two or three weeks. It then passes into a putrefying decomposition, acquires the odour of decaying cheese, and, like decaying cheese, has now the property of changing sugar inte *lactic acid*, instead of into alcohol, as before. A hundred pounds of crushed grain will yield six to eight pounds of the crushed yeast. It is made largely at Rotterdam, and is imported thence to this country through Hull.

refined physiological chemistry, how, from the same food taken in from the air, and from generally similar food drawn up from the soil, different plants, and different parts of plants, should be able to extract or produce substances so very different from each other in composition and in all their properties. From the seed-vessels of one (the poppy), we collect a juice which dries up into our commercial opium; from the bark of another (cinchona) we extract the quinine with which we assuage the raging fever; from the leaves of others, like those of hemlock and tobacco, we distil deadly poisons, often of rare value for their medicinal uses. The flowers and leaves and seeds of some yield volatile oils, which we delight in for their odours and their aromatic qualities; the seeds of others give fixed oils, which are prized for the table or for use in the arts. The wood of some is rich in valuable dyes, while from that of others exude turpentines and resins of varied degrees of worth—from the cheap rosin of the tinsmith and soapmaker to the costlier myrrh and aloes and benzoin which millions still burn, as acceptable incense, before the altars of their gods.

These, and a thousand other similar facts, tell us how wonderfully varied are the changes which the same original forms of matter undergo in the interior of living plants. Indeed, whether we regard the vegetable as a whole or examine its minutest parts, we find equal evidence of the same diversity of changes, and of the same production, in comparatively minute quantities, of very different, yet often very characteristic forms of matter.

Thus, looking at a large tree as a whole, we are charmed with the brilliant green foliage which invests it when summer has come, and to which the landscape owes half its charms. Yet chemistry tells us that all this effect of colour is produced by the fraction of an ounce of colouring matter distributed evenly over its thousands of leaves! Or taking up the leaf of a nettle, and picking off one of its minute stinging prickles, chemistry, by the aid of the microscope, assures us that the pain it causes, when allowed to pierce the skin, arises from a reservoir of a peculiar acid (the formic acid), which, like the poison of the serpent's tooth, is squeezed into the wound which the spikelet makes.

Fig. 16.

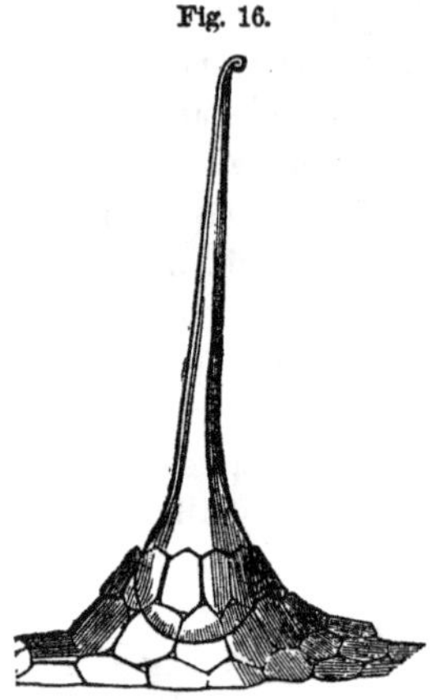

The acid is contained in these elastic cells at the base of the prickly hair.

The characteristic property of the minute nettle-hair, and the peculiar charm of the wide landscape, are equally dependent upon the production in living plants of special forms of matter in comparatively minute proportions.

The tuber of the potato, the ripening apple, and the growing twig, present us with another illustration of special chemical changes proceeding continuously in the plant, and with a definite reference to a specific and useful end. The unripe potato, when taken from the earth, withers and shrivels, becomes unsightly to the eye, and vapid to the taste; the unripe apple shrinks in, refuses to retain its natural dimensions, and cannot be kept for any length of time; while the unripe twig perishes amidst the chills of winter, and remains black and dead when the green buds of spring were expected to enliven its surface. These effects are the consequence of the thin bark which covers potato, apple, and twig alike, not having attained its matured composition. While unripe, this coating is porous and pervious to water, so that, when removed from the parent plant, tuber, fruit, and twig all give off water by evaporation to the air, and thus shrivel and shrink in as I have described. But when ripe, this porous covering has become chemically changed into a thin impervious coating of *cork*, through which water can scarcely pass, and by which, therefore, it is confined within for months together. It is this cork-layer which enables the potato to keep the winter through, the winter pear and winter apple to be brought to table in spring of their full natural dimensions, and the ripened twig to retain its sap undried, and to feed the young bud when the April sun first wakens it from its winter's sleep.

Nor are the general purposes for which the entire plant lives, and is the theatre, so to speak, of so many changes, to be properly, I may say at all appreciated without the assistance of chemical research.

It is true that every one can recognise in the natural herbage and the wild forest the ornaments of the landscape; in the thousand odours they distil, and in the varied hues and forms with which they sprinkle the surface, the most agreeable and refined ministers to our sensual pleasures. And in these things we unquestionably see some of the true purposes served by vegetation in the economy of nature. But they are subsidiary purposes—which they serve, by the way, as it were, while labouring to fulfil their true and greater vocation.

This vocation may be viewed in two aspects—*first*, as regards dead nature; and, *second*, as regards living things.

First, In its relations to dead nature, the plant serves, while living, to purify the air we breathe. It continually absorbs carbonic acid and gives off oxygen gas, and thus is a chief instrument in maintaining the normal condition of the atmosphere. It renders the air more fit for the support of animal life, both by removing that which is noxious (the carbonic acid), and by pouring into it that which is salutary (the oxygen) to animal health and life. And then, when it dies, it either covers the earth with a vegetable mould, which favours the growth of new generations of plants, or it accumulates into beds of peat or mineral coal, by which man is long after to be warmed, and the arts of life promoted. But in either case it only lingers for a while in these less sightly mineral forms. It gradually assumes again the gaseous state, and whether it is allowed naturally to decay, or is burned in the fire, ultimately arises again into the air in the form of carbonic acid. By this means, in part, vegetation is perpetuated upon the globe, and the natural composition of the atmosphere, as regards the proportion of the carbonic acid gas, is permanently maintained. And,

Second, As regards living animals, we all know and feel that plants are necessary to our daily life. Utterly dry up and banish vegetation from a region, and nearly every sensible form of animal life forthwith disappears. But how do plants feed us? And by what virtues in their several parts can the ox thrive on the straw, while man can live only on the grain? How on the nut and fruit of the tree only can human life be permanently sustained, while the leaves and twigs of the thick forest sustain the lordly elephant?

As to dead nature, the plant serves a subsidiary purpose in covering and adorning it—so to living nature, to man especially, it serves a similar subsidiary purpose in producing the numerous remarkable products, to which I have already alluded as being useful in medicine and the arts, and as ministering to the luxuries and comfort of civilised life. In the production of these we recognise a destined and benevolent purpose served by the general vegetation of the globe, in reference to living things. But this purpose is only secondary, and, as it were, ornamental. The main object of the plant, in its relations to the animal, is to feed it. This it does with various forms of vegetable matter in different climes and countries, and it provides for each herbivorous and carnivorous race those peculiar forms on which it best loves, because best fitted, to feed. It is so with man His vegetable food varies with the part of the world in which he is situated; yet upon all the varieties with which different climates furnish him, he discovers the means continuously to sustain himself.

Of what chemical substances do these different forms of nutritious food consist? What do they possess in common? In what do they differ? Why do some of them, weight for weight, sustain the body more completely or for a longer time than others? Why do they affect the dispositions of those who consume them—not only the constitution of individuals, but the habits, temperament, and character of whole nations? Why do we choose to mix the forms of vegetable food we consume—whence come the fashions of universal cookery—whence the peculiarities of national dishes?

What a host of curious chemical inquiries spring up in connection with the plant we rear, regarded as the main sustenance or staff of common life? I shall consider some of them in the following chapter.

CHAPTER V.

THE BREAD WE EAT.

The grain of wheat.—Bran and flour.—Separation of flour into starch and gluten.—Fermenting of dough.—Baking of bread.—New and stale bread.—Proportion of water in flour and in bread.—Composition of bread.—Bran richer in gluten.—Comparative composition.—Wheaten and rye bread compared.—Oatmeal and Indian-corn meal.—Composition of rice.—Buckwheat, quinoa, Guinea corn, and dhurra.—Composition of beans, peas, and lupins.—The sago palm, and the seeds of the araucaria.—The fruits of the banana, the date palm, the fig tree, and the bread-fruit tree.—Water contained in fruits and roots.—The turnip, carrot, and potato.—The composition of rice, the potato, and the plantain compared.—Deformity among the eaters of these three vegetables.—The Siberian lily.—The use of leaves as food.—The cabbage very nutritious.—Natural tendency of man to adjust the constituents of his food.—Irish kol-cannon.—Starvation upon arrow-root and tapioca.—General character of a nutritious diet.—National and individual influence of diet.

THE bread we eat I take as the type of our vegetable food. On such food of various kinds, and eaten in various forms, man and animals are sustained in all parts of the globe. The study of our common wheaten bread will give us the key to the composition and known usefulness of them all.

1°. WHEAT.—When the grain of wheat is crushed between the stones of the mill, and is then sifted, it is separated into two parts—the bran and the flour. The bran is the outside, harder part of the grain, which does not crush so readily, and when it does crush, darkens the colour of the flour. It is therefore generally sifted out by the miller, and is used for feeding horses, pigs, and other animals, or even for applying to the land as a manure.

If the flour be mixed with a quantity of water sufficient to moisten it thoroughly, the particles cohere and form a smooth, elastic and tenacious dough, which admits of being drawn out to some extent, and of being moulded into a variety of forms. If this dough be placed upon a sieve or on a piece of muslin, and worked with the hand under a stream of water (fig. 17), as long as the water passes through milky, there will remain at last upon the sieve a white sticky substance, very much resembling birdlime. This is the substance which gives its tenacity to the dough. From its glutinous character it has obtained among chemists the name of gluten. When the milky water has become clear by standing, a white powder will be found at the bottom of the vessel, which is common wheaten starch. Thus the flour of wheat contains two principal substances, gluten and starch. Of the former, every 100 lb. of fine English flour contain about 10 lb., and of the latter about 70 lb.

Fig. 17.

Mode of separating the gluten from the starch of Wheat.

The way in which the bran, the gluten, and the starch are respectively distributed throughout the body of the seeds of our corn plants is shown in the following section of a grain of rye when fully ripe.

In the figure to the left, *a* represents the outer seed-coat, consisting of three rows of thick-walled cells; *b* the inner

Fig. 13.

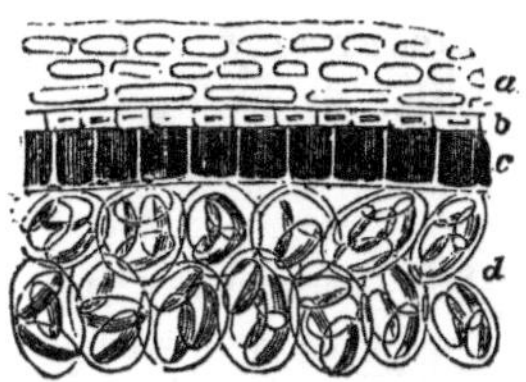

Structure of a grain of Rye.

Single cell.

Grains of Rye Starch, (see p. 100).

seed-coat, composed of a single layer of thick-walled cells, having scarcely any cavity; *c* a layer of cells containing gluten. These three together form the bran. *d* represents the cellular tissue of the albumen,* consisting of large roundish hexagonal cells, which contain grains of starch.

The middle figure exhibits one of the cells of the albumen more highly magnified, and shows how the grains of starch are disposed in it. The small figures to the right are grains of starch still more highly magnified. Their natural size varies from a ten-thousandth to a six-hundredth of an inch.

The outer coating contains only three or four per cent. of gluten, the inner coating from fourteen to twenty per cent. All this is separated in the bran. Throughout the mass of the grain around and within the albumen cells the gluten is diffused every where among the grains of starch.

When a little yeast is added to the flour before or while it is being mixed with water into a dough, and the dough is then placed for an hour or two in a warm atmosphere, it begins to *rise*—it ferments, that is, and swells or increases in bulk. Bubbles of gas (carbonic acid gas) are disengaged

* The reader must not confound this word *albumen*, used by botanists to denote the white inner part of the seed, with the same word used in chemistry as the name of the *white of the egg*.

in the interior of the dough, which is thereby rendered light and porous. If it be now put into a hot oven, the fermentation and swelling are at first increased by the higher temperature; but when the whole has been heated nearly to the temperature of boiling water, the fermentation is suddenly arrested, and the mass is fixed by the after baking in the form it has then attained.*

It is now newly-baked bread, and if it be cut across it will appear light and spongy, being regularly sprinkled over with little cavities, which were produced in the soft dough by the bubbles of gas given off during the fermentation. This fermentation is the consequence of a peculiar action, which yeast exercises upon moist flour. It first changes a part of the starch of the flour into sugar, and then converts this sugar into alcohol and carbonic acid, in the same way as it does when it is added to the worts of the brewer or the distiller. As the gas cannot escape from the glutinous dough, it collects within it in large bubbles, and makes it swell, till the heat of the oven kills the yeast plant, and causes the fermentation to cease. The alcohol escapes, for the most part, during the baking of the loaf, and is dissipated in the oven.

New-baked bread possesses a peculiar softness and tenacity which is familiar to most people, and though generally considered less digestible is a favourite with many. After two or three days it loses its softness, becomes free and crumbly, and apparently drier. In common language, the bread becomes stale, or it is stale bread. It is generally supposed that this change arises from the bread becoming actually drier by the gradual loss of water; but this is not the case. Stale bread contains almost exactly the same proportion of water as new bread after it has become completely

* The formation of hard crusts on the loaf may be prevented by rubbing a little melted lard over it after it is shaped, and before it is set down to rise, or by baking it in a covered tin.

cold. The change is merely in the internal arrangement of the molecules of the bread. A proof of this is, that if we put a stale loaf into a closely-covered tin, expose it for half an hour or an hour to a heat, not exceeding that of boiling water, and then remove the tin, and allow it to cool, the loaf, when taken out, will be restored in appearance and properties to the state of new bread.

The quantity of water which well-baked wheaten bread contains, amounts on an average to about forty five per cent. The bread we eat, therefore, is nearly one-half water;—it is, in fact, both meat and drink together.

The flour of wheat and of other kinds of grain contains water naturally, but it absorbs much more during the process of conversion into bread. One hundred pounds of fine wheaten flour take up fifty pounds, or half their weight of water, and give 150 pounds of bread. Thus, 100 of English flour and 150 of bread contain respectively—

	The flour contains	The bread contains
Dry flour, . . .	84	84
Natural water, . .	16	16
Water added, . . .	....	50
	100 lb.	150 lb.

One of the reasons why bread retains so much water is, that during the baking a portion of the starch is converted into gum, which holds water more strongly than starch does. A second is, that the gluten of flour, when once thoroughly wet, is very difficult to dry again, and that it forms a tenacious coating round every little hollow cell in the bread, which coating does not readily allow the gas contained in the cell to escape, or the water to dry up and pass off in vapour; and a third reason is, that the dry crust which forms round the bread in baking is nearly impervious to water, and, like the skin of a potato which we bake in the oven or in the hot cinders, prevents the moisture within from escaping.

The proportions of water, gluten, and starch or gum, in well-baked wheaten bread, are nearly as follows:—

Water, ...	45
Gluten, ...	6
Starch, sugar, and gum, ...	49
	100

The bran or husk of wheat, which is separated from the fine flour in the mill, and is often condemned to humbler uses, is somewhat more nutritious than either the grain as a whole, or the whiter part of the flour. The nutritive quality of any variety of grain depends very much upon the proportion of gluten which it contains; and the proportions of this in the whole grain, the bran and the fine flour respectively, of the same sample of wheat, are very nearly as follows:

Whole grain, ...	12 per cent.
Whole bran (outer and inner skins).	14 to 18 "
Fine flour, ...	10 "

If the grain, as a whole, contain more than twelve per cent. of gluten, the bran and the flour will also contain more than is above represented, and in a like proportion. The *whole meal* obtained by simply grinding the grain is equally nutritious with the grain itself. By sifting out the bran we render the meal less nutritious, weight for weight; and when we consider that the bran is rarely less, and is sometimes considerably more, than one-fourth of the whole weight of the grain, we must see that the total separation of the covering of the grain causes much waste of wholesome human food. Bread made from the whole meal is therefore more nutritious; and as many persons find it also a more salutary food than white bread, it ought to be more generally preferred and used.

The woodcut and explanations given in p. 81, show that the gluten of the husk resides chiefly in the inner covering of the grain. Hence the outer covering may be removed without sensible loss of nutriment, leaving the remainder both more nutritious than before, weight for weight, and also more digestible than when the thin outer covering is left upon the corn. An ingenious American instrument has been patented, by which this removal of the outer coating is said to be completely effected without injury to the bulk of the grain.

It is also a point of some interest that the small or tail corn, which the farmer separates before bringing his grain to market, and usually grinds for his own use, is richer in gluten than the plump full-grown grain, and is therefore more nutritious.

2°. Barley and Rye resemble the grain of wheat very much in composition and nutritive quality. They differ from it somewhat in flavour and colour, and do not make so fair and spongy a bread. They are not generally preferred, therefore, in countries where other grains thrive and ripen. Two samples of newly-baked wheat and rye bread, made and examined under the same circumstances, were found to consist respectively of

	Wheaten bread.	Rye bread.
Wator,	48	48½
Gluten,	5¾	5⅓
Starch, &c.,	46¼	46⅙
	100	100

So that in composition and nutritive quality these two kinds of bread very closely resemble each other; and except as concerns our taste, it is a matter of indifference whether we live on the one or the other. Rye bread possesses one quality which is in some respects a valuable one: it retains its freshness and moisture for a longer time than wheaten bread, and can be kept for months without becoming hard, dry, or unpalatable. This arises principally from certain peculiar properties possessed by the variety of gluten which exists in the grain of rye.

3°. INDIAN CORN also resembles wheat in composition and nutritive quality. Its grain has a peculiar flinty hardness, and its flour, usually known as Indian meal, a flavour which in this country is not at first relished. It does not bake into the same light spongy loaves as wheaten flour, but it is excellent in the form of cakes. The chief peculiarity in its composition is, that it contains more oil or fat than any of our common grains. This oil sometimes amounts to as much as nine pounds in the hundred, and is supposed to impart to Indian corn a peculiar fattening quality.

4°. OATS are a favourite food in our island for horses, and in Scotland especially are much esteemed as an agreeable, nutritious, and wholesome food for man. The meal of this grain is distinguished for its richness in gluten, and for containing more fatty matter than any other of our cereal grains. To these two circumstances it owes its eminently nutritious and wholesome character. The average relative proportions of gluten, fat, and starch contained in fine wheaten flour, in Scotch oatmeal, and in Indian-corn meal, are represented by the following numbers :—

	English fine wheaten flour.	Bran of English wheat.	Scotch oatmeal.	Indian-corn meal.
Water, .	16	13	14	14
Gluten, .	10	18	18	12
Fat, . . .	2	6	6	8
Starch, &c.,	72	63	62	66
	100	100	100	100

The large proportion of fatty matter contained in Indian corn not only adapts it well for fattening animals, but makes it more grateful to the alimentary canal, and therefore more wholesome. I have inserted in the above table a column showing the average composition of the bran of English wheat, for the purpose of showing, *first*, how large a proportion of fat it also contains, compared with fine wheaten flour; and, *second*, the remarkable similarity in composition, in some respects, which exists between the bran of wheat and the meal of the oat.

Owing to a peculiar quality of the gluten which the oat contains, the meal of this grain does not admit of being baked into a light fermented spongy bread. It has been alleged against oatmeal, that when used as the sole food, without milk or other animal diet, it produces heat and irritability of the skin, aggravates skin diseases, and sometimes occasions boils, in the same way as salt meat tends to produce scurvy. Dr. Pereira, a high authority, says that this charge has been made without just grounds. At all events, it must be very rarely that circumstances render necessary for any length of time such an exclusive consumption of oatmeal.

5°. Rice is remarkable chiefly for the comparatively small proportion of gluten it contains. This does not exceed seven or eight per cent.—less than half the quantity contained in oatmeal. In rice countries it has often been noticed, that the natives devour what to us appear enormous quantities of the grain, and this circumstance is ascribed to the small proportion it contains of the highly nutritive and necessary gluten. Rice contains also little fat, and hence it is less laxative than the other cereal grains, or rather it possesses something of a binding quality. It has recently been observed that, when substituted for potatoes in some of our workhouses—in consequence of the failure of the potato—

this grain has after a few months produced scurvy. This may have been owing as much to the effects of sudden change of diet as to an inherent evil property in the grain itself. Still it suggests, as many other facts do, the utility and wholesomeness of a mixed food.

6°. Buckwheat flour is about as nutritious as English wheaten flour, and makes excellent cakes, which, when eaten hot with maple honey, in the backwoods of America, are really delicious.

7°. Quinoa.—A variety of grain scarcely known in this country is the quinoa (fig. 19), a small roundish seed, which is extensively cultivated and consumed on the high table lands of Chili and Peru. There are two varieties of it—the sweet and the bitter—and both grow at elevations rising to 13,000 feet above the level of the sea, where both rye and barley refuse to ripen. It is still the principal food of the many thousands of people who occupy these high lands, and, before the introduction of European grains by the Spaniards, is said to have formed the chief nourishment of the Peruvian nation. It is very nutritious, and in its composition approaches very nearly that of oatmeal. Thus the flour or meal of the oat and of the quinoa consist respectively of—

Fig. 19.

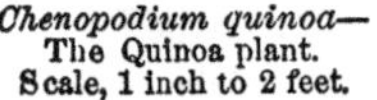

Chenopodium quinoa—The Quinoa plant. Scale, 1 inch to 2 feet.

	Oatmeal.	Quinoa flour. (Voelcker.)
Water,	14	16
Gluten,	18	19
Fat,	6	5
Starch, &c.,	62	60
	100	100

A grain so nutritious as this is a very precious gift to the inhabitants of the elevated regions of the Andes. Without it, these lofty plains could only be runs for cattle, like the summer pastures among the valleys on the Alps.

8°. GUINEA CORN, a small seed, used to some extent in the West Indies, is a little less nutritious than ordinary English wheat.

9°. DHURRA or DHOORA (fig. 20), a small kind of grain much cultivated and extensively consumed in India, Egypt, and the interior of Africa, is quite equal in nutritive value to the average of our English wheats, and yields a beautiful white flour. According to my analysis, buckwheat flour contains 10½, and dhurra flour 11½ per cent. of gluten.

10°. The BEAN, the PEA, the LUPIN, the VETCH, the LENTIL, and other varieties of pulse, contain, as a distinguishing character of the whole class, a large per-centage of gluten, mixed with a comparatively small per-centage of fat. On an average, the proportion of gluten is about twenty-four, and of fat about two in every hundred. The gluten of these kinds of grain resembles that of the oat, and does not, therefore, fit bean or pease meal for being converted into a spongy bread. The large proportion in which this ingredient is present in them, however, renders all kinds of pulse very nutritious. Eaten alone, however, they have a constipating or costive quality; but a proper admixture of them with other kinds of food, especially

Fig. 20.

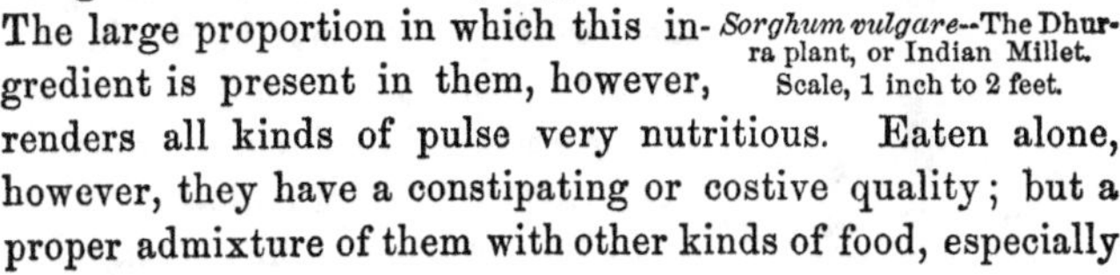

Sorghum vulgare--The Dhurra plant, or Indian Millet. Scale, 1 inch to 2 feet.

with such as contain a larger proportion of oil or fat, is found to give both strength and endurance to animals which are subjected to hard labour. It is in this way that a certain quantity of beans given to horses among their oats, is found so serviceable in this country.

It is because also of the same large per-centage of gluten that the chick pea, the *gram* of the East, is considered, when roasted, to be more capable of sustaining life, weight for weight, than any other kind of food. For this reason it is selected by travellers about to cross the deserts, where heavy and bulky food would be inconvenient.

Of all these varieties of grain a kind of bread is made by those who live upon them, and they are all more or less used in this form for human food. Only two of them, however, I believe—wheat and rye—possess the property, when mixed with yeast or leaven, of forming a light spongy bread, which cannot be kept for a time without becoming unpalatable. And of the two varieties of bread yielded by these grains, that made from wheat is the more dry and crumbly, the more fair to look upon, and the more agreeable to the taste. Hence the universal preference which exists for the flour of wheat and for wheaten bread wherever they can easily be obtained.

Fig. 21.

Sagus rumphii—The Sago Palm.
Scale, 1 inch to 20 feet.

But trees also share with corn-bread to a considerable extent in the nutrition of the human race. Among these, the sago palm, the Chilian pine, the banana or plantain, and the date, the fig and the bread-fruit tree, are deserving of especial notice.

11°. The SAGO PALM (*Sagus

rumphii) is cultivated in many places, but it is the chief support of the inhabitants of north-western New Guinea, and of parts of the coast of Africa. The meal is extracted from the pith by rubbing it to powder, and then washing it with water upon a sieve. It is baked by the natives into a kind of bread or hard cake by putting it for a few minutes into a hot mould. The exact nutritive value of this meal has not been chemically ascertained. It has been stated, however, that 2½ lb. of it are sufficient to serve for a day's

Fig. 22.

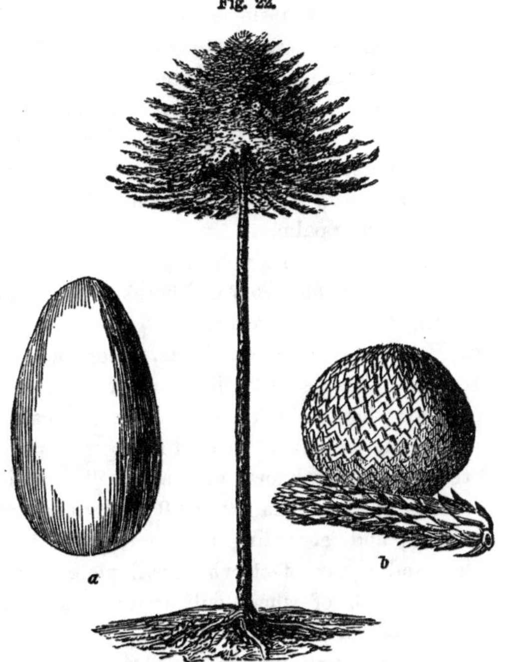

Araucaria imbricata—The Chili Pine.
Scale, 1 inch to 40 feet.
a Kernel of seed, the natural size—*b* Cone, 1 inch to 10 inches.

sustenance to a healthy full-grown man. And as each tree, when cut down in its seventh year, yields seven hundred pounds of sago meal, it has been calculated that a single acre of land planted with three hundred trees—one-seventh to be cut down every year—will maintain fourteen men.

12°. But the CHILIAN PINE (*Araucaria imbricata*), now known among us for its beauty, is still more conspicuous as a feeder of men. In our British woods the tiny squirrel supports its life during the winter months on the seeds of the larch, the pine, and the Scotch fir, which we plant for ornament or use. But on the slopes of the Andes of Chili and Patagonia the lofty araucaria extends in natural forests, bearing huge cones six inches in diameter. The seeds contained in these are large, and supply the natives with a great part of their usual food. "The fruit of one large tree will maintain eighteen persons for a year;" and this, year by year, without the necessity of cutting down and replanting, as in the case of the sago palm.

We do not know the composition of pine seeds, but they probably do not differ much from the beech-nut, the chestnut, and the acorn, all of which are rich in gluten.

13°. THE BANANA.—Of some fruits, tales nearly as wonderful are told. The beautiful banana, for example, the ornament of country-houses in tropical countries, is said to yield from the same extent of ground a larger supply of human food than any other known vegetable. The fruit of a single tree sometimes weighs 70 or 80 lb., but averages from 30 to 40 lb.; and, according to Humboldt, the same space of a thousand square feet, which will yield only 462 lb. of potatoes, or 38 lb. of wheat, will produce 4000 lb. of bananas, and in a shorter period of time!

The fruit, however, contains seventy-three per cent. of water. Even when dried and converted into meal, it is less nutritious than the meal of any of the varieties of grain above mentioned. It approaches in composition most nearly to rice, and, like that grain, swells out the stomachs of the negroes who feed upon it. In tropical countries it is nevertheless a most valuable food, and is so extensively consumed as to take the place of our cereal grains as the common article of diet. About 6½ lb. of the fruit, or 2 lb. of the dry meal, with ¼ lb. of salt meat or fish, form, in tropical America, the daily allowance for a labourer, whether slave or free.

Fig. 23.

Musa sapientum—The Banana Tree.
Scale, 1 inch to 10 feet.
Fruit, 1 inch to 5 inches.

The unripe fruit is sometimes used as bread; it is dried in the oven, and in this state is eaten in the manner of bread. When thus dried, it may be kept for a long time without spoiling, and is usually carried with them in this dry state by the natives when they are proceeding on a long journey.

The chemical reason why the unripe fruit is chosen for this purpose, is, that while unripe, the fruit is filled with starch, so that when dried it has a resemblance to bread both in taste and composition. As the fruit ripens, this starch changes into sugar, and the fruit becomes sweet. In this state, though more pleasant to eat when newly pulled, it is less fit either for drying or for preserving.

Fig. 24.

Phœnix dactylifera--The Date Palm.
Scale, 1 inch to 20 feet.
Fruit, 1 inch to 2 inches.

14°. THE DATE.—Many other fruits are more nutritious, weight for weight, than the banana, though none may probably be compared with it as an abundant producer of food. The date, for example, "the bread of the desert," is capable of supporting life, and of sustaining unaided the strength of man, for an indefinite period. We possess no chemical results from which to judge of the absolute nutritive quality of this fruit, but experience proves that it must be very considerable.

The date palm (*Phœnix dactylifera*), the tree which yields this fruit, is invaluable amid parched sands and arid deserts. Wherever a spring of water appears amid the sandy deserts of Africa (between 19° and 35° N. latitude), this graceful palm yields at once both its grateful shelter and its nourishing fruit. Where all other crops fail from drought, the date tree still flourishes. In Egypt and Arabia it forms a large portion of the general food, and among the oases of Fezzan "nineteen-twentieths of the population live upon it for nine months in the year."

15°. THE FIG.—The fig, like the date, is a native of warm climates. Of the chemical history of this and some other fruits we know more than we do as yet of the date. In the perfectly dry state it is about as nutritious as rice. In the moist state, as it is imported, it will go considerably further in feeding, and especially in fattening or adding generally to the weight of an animal, than an equal weight of wheaten bread!

Thus, figs as imported, and wheaten bread in its usual state, consist respectively of—

	Figs.	Wheaten bread.
Water, ...	21	48
Gluten, ...	6	5¾
Starch, sugar, &c., ...	73	46¼
	100	100

The fig, it will be seen by comparing the above columns, contains about as much gluten as wheaten bread, while in starch and sugar it is twenty-seven per cent. richer. The perfectly dry gooseberry is about as nutritive as ordinary wheaten flour.

16°. THE BREAD-FRUIT TREE (*Artocarpus incisa*) is remarkable for its large and brilliant leaf, and for the general beauty of its appearance, in which respect none of our forest trees can compare with it. But it is most remarkable for the abundant, peculiar, and nutritious fruit it yields. This fruit is nearly round, and attains to a considerable size. It grows abundantly, and covers the tree for eight or nine months without interruption, and the crops ripen in succession. There are various ways of cooking it, for it is seldom relished raw. While the fruit is on the tree, it is plucked before it is perfectly ripe, while the rind is still green, but the pith snow-white, and of a porous and mealy texture. It is then peeled, wrapped in leaves, and baked on hot stones. In this state it tastes like wheaten bread, sometimes rather sweeter. When quite ripe, the starch, as in the banana, has become partly changed into sugar, so that the pith is pulpy and of a yellow colour, and can be eaten uncooked, but it has still a disagreeable flavour. To serve for food during the three months when the tree ceases to bear, the unripe fruits, after being peeled, are laid in a paved pit and covered with leaves and stones; they there ferment and become sour, and form a kind of paste, which tastes like black Westphalian bread when not thoroughly baked. The quantity required for daily use is taken from the pit, made into lumps about the size of the fist, rolled in leaves, and baked on stones as before. These lumps of bread keep for weeks, and are a very good provision in journeys.

Fig. 25.

Artocarpus incisa—The Bread-fruit Tree.
Scale, 1 inch to 40 feet.
Leaf and fruit, 1 inch to a foot and a half.

The crops of this fruit are so abundant that three trees are sufficient to maintain a man for eight months. It is more productive, therefore, even than the banana or the sago tree. "Whoever," says Captain Cook, "has planted ten bread-fruit trees, has fulfilled his duty to his own and succeeding generations as completely and amply as an inhabitant of our rude clime who, throughout his whole life, has ploughed during the rigour of winter, reaped in the heat of summer, and not only provided his present household with bread, but painfully saved some money for his children."

On the islands of the Indian Archipelago, and on the island groups of the South Sea, this tree is found. The fruit is best, however, on the Friendly and Marquesas Islands. It has never been observed wild, but the whole species has passed into a cultivated state, and it is therefore probable, says Meyen, "that man settled wherever he found a bread-fruit tree. Even yet the favourite situation of the fragile Indian huts is under its shady branches." *

The chemical composition of this fruit has never been determined. We know by its properties that, while unripe, it contains much starch, which during the ripening is partly changed into sugar; but how much gluten or oily matter, or even of water, is present in it, has not, so far as I am aware, been hitherto experimentally ascertained.

The quantity of water they contain is a character of fruits which is very important. By this they are distinguished in a remarkable manner from the different varieties of grain. Thus the fruit of the

Plantains contains .	73	per cent. of water.
Plums, and other fleshy fruits,	75	" "
Apples, gooseberries, &c.,	80	" "

The consequence of this composition is, that in fruits all the nutritive matter is diluted with a large quantity of water, and in this state experience has shown that all nutritive substances are more grateful to the healthy stomach and more easily digested. It is for this reason that, in preparing our

* MEYEN'S *Geography of Plants* (Ray Society), p. 321.

dry grains for food, we almost invariably imitate this preparatory process of nature. Even in baking our bread, as we have seen above, the result of our operations is that we convert it into a light and spongy mass containing nearly half its weight of water. And yet we talk of this as *dry* bread, and rarely eat it without some accompanying fluid.

The ROOTS and TUBERS we use as food occur naturally in the same watery condition as fruits do. The potato, the carrot, and the turnip, for example, contain respectively in a hundred pounds—

	Water.	Dry food.
The potato, ...	75	25
" carrot, ...	83	17
" turnip, ...	90	10

The gourd tribe are still more remarkable for the quantity of water they contain. The water melon, for example, contains ninety-four per cent., and the cucumber ninety-seven per cent. of water! No wonder that Jonah's gourd could spring up in a night—that this tribe of plants should be so much esteemed in hot climates, where thirst rages—or that old Mehemet Ali should have been able to eat up an entire forty pound melon after the substantials of his dinner were disposed of!

17°. THE TURNIP AND CARROT.—The dry substance of the roots and green vegetables we use as food resembles that of seeds and fruits in general composition. The dried meal of the turnip and carrot, for example, contains gluten associated with starch and sugar, and is very nutritious. That of the turnip is quite equal in this respect to Indian-corn meal, being only deficient in fat. Hence a little oily food should be always used along with a turnip diet. Attempts have been made to manufacture a palatable meal from dried turnips, but the disagreeable taste of the root so clings to the meal as hitherto to have rendered it unsuited for human consumption.

18°. THE POTATO is more important as a variety of human food than any other root we cultivate, and is remarkable for being grown over a greater range of latitude than any other cultivated plant. The dry substance which it contains—the potato meal, that is—is unsuited for being made into bread alone, though it is used to some extent, as an admixture with wheaten flour, and is said in most cases to improve the bread in lightness and general appearance. The dried potato is less nutritive, weight for weight, in the sense of supporting the strength, and enabling a man to undergo fatigue, than any other extensively-used vegetable food, of which the composition is known, with the exception only of rice and of the plantain. It approaches nearest, indeed, to rice, though it is somewhat superior to that grain. Thus, the dry substance of these three forms of food consists of—

	Rice.	Potato.	Plantain.
Gluten,	7½	8	5¼
Starch, &c.,	92½	92	94¾
	100	100	100

There is, therefore, a remarkable similarity among these three kinds of food, in so far as they all differ from our cereal and other grains and roots, in containing a smaller proportion of the ingredient represented by the gluten of wheat. And in the use of them all, it is remarkable that a chemical or physiological likeness is indicated by the observation that the tribes of people who live exclusively or even chiefly on any of these three vegetable productions, are distinguished by the size and prominence of their stomachs! The Hindoo who lives on rice, the negro who lives on the plantain, and the Irishman who lives exclusively on the potato, are all described as being more or less pot-bellied. This peculiarity is to be ascribed in part, I suppose, to the necessity of eating a large bulk of food, in order to be able to extract from it a sufficient amount of necessary sustenance. And that this deformity is somewhat less conspicuous in the Irish potato-eater than in the plantain-loving negro, or even the rice-devouring Chinaman and Hindoo, is probably to be ascribed to the somewhat larger proportion of the gluten ingredient which is present in the potato.

One remarkable circumstance in which the three kinds of meal just spoken of differ from each other, is in the size of the grains of starch in each. As seen in the following figures—all drawn to the same scale—the starch granules in the potato are very large, having sometimes a length of two or three thousandths of an inch. Those of the plantain, though considerably larger than the granules of wheat or rye (p. 81), average less than half the size of those of the potato; while those of rice are angular, and have an average diameter of less than one five-thousandth of an inch.

Fig. 26.

a Granules of Potato Starch. *b* Granules of Plantain Starch. *c* Granules of Rice Starch.

Whether the peculiarities above shown influence in any

way the nutritive action of these several kinds of food, has not yet been specially investigated.

19°. THE ONION is worthy of notice as an extensive article of consumption in this country. It is largely cultivated at home, and is imported, to the extent of seven or eight hundred tons a-year, from Spain and Portugal. But it rises in importance when we consider that in these latter countries it forms one of the common and universal supports of life. It is interesting, therefore, to know that in addition to the peculiar flavour which first recommends it, the onion is remarkably nutritious. According to my analyses, the dried onion root contains from twenty-five to thirty per cent of gluten. It ranks in this respect with the nutritious pea and the *gram* of the East. It is not merely as a relish, therefore, that the wayfaring Spaniard eats his onion with his humble crust of bread, as he sits by the refreshing spring: it is because experience has long proved that, like the cheese of the English laborer, it helps to sustain his strength also, and adds—beyond what its bulk would suggest—to the amount of nourishment which his simple meal supplies.

20°. Among roots which are important articles of diet in more limited districts, may also be mentioned the tuber of a lily (*Lilium pomponium*) which is roasted and eaten in Kamtschatka, and is there cultivated as we do the potato. That it is nutritious is certain, but with its exact chemical composition and nutritive value we are as yet unacquainted.

LEAVES.—From roots we turn to leaves, which form no inconsiderable proportion of the daily sustenance of European nations. The greater number of animals, wild as well as domestic, live upon the leaves of plants. Our oxen feed upon the grasses; and even the huge elephant and the sloth find their nourishment on the leaves of the forests in which they live. Among those which are raised for human food, the cabbage is a regular field crop; and many others are cultivated less extensively in our gardens.

Leaves are generally rich in gluten; many of them, however, contain other substances in smaller quantity associated with the gluten, which are unpleasant to the taste, or act injuriously upon the general health, and therefore render them unfit for human food. Dried tea-leaves, for example, contain about twenty-five per cent of gluten; and therefore, if they could be eaten with relish, and digested readily, they would prove as strengthening as beans or peas.

21°. THE CABBAGE is an especially nutritious vegetable. The dried leaf contains, according to my analyses, from thirty to thirty-five per cent. of gluten, and is, in this respect, therefore, more nutritious than any other vegetable food which is consumed to a large extent by men and animals. I know, indeed, of only two exceptions—the mushroom, which in its dry matter contains sometimes as much as fifty-six per cent. of gluten—and the dried cauliflower, in which the gluten occasionally rises as high as sixty-four per cent.

The cabbage is one of those plants from the leaves of which, by boiling, we can extract the greater part of that which is disagreeable to the taste, and thus convert it into a palatable food, without sensibly diminishing its nutritious quality.* When eaten frequently, however, and in large quantity, they have, in common with nearly all kinds of food which are rich in gluten, a costive or binding tendency upon the human constitution; hence the propriety of eating them with fat and oily food. Bacon and greens, like pork and pease-pudding, is a conjunction of viands which does not owe its popularity either to old habit or to the mere

* I found, for example, that the dried matter of boiled cabbage still contained thirty-three per cent. of gluten.

taste of the epicure. It is in reality an admixture which constitutional experience has prescribed as better fitted to the after comfort of the alimentary canal of every healthy individual, than either kind of food eaten alone.

And so with a dish common in Ireland under the name of Kol-cannon. The potato, as we have seen, is poor in gluten—the cabbage is unusually rich in this ingredient; mix the two, and you approach the composition of wheaten bread. Beat the potatoes and boiled cabbage together, put in a little pork fat, salt, and pepper, and you have a kol-cannon which has all the good qualities of the best Scotch oatmeal, and to many would be more savoury and palatable. Take a pot-bellied potato-eater, and feed him on this dish, and he will become not only stronger and more active, but he will cease to carry before him an advertisement of the kind of food he lives upon, and his stomach will fall to the dimensions of the same organ in other men.

Such are the principal varieties of vegetable food which —partly in the form of baked bread, and partly cooked in other ways—are at the present day most largely employed in the feeding of the human race. We have seen in all of them—

First, That they contain a sensible proportion of three important constituents—gluten, starch, and fat.

Second, That when the proportion of any of these is too small, chemistry indicates, and experience suggests, that an additional quantity of this deficient substance should be added in the process of cooking, or preparatory to eating. Thus we consume butter with our bread, and mix it with our pastry, because wheaten flour is deficient in natural fat; or we eat cheese or onions with the bread, to add to the proportion of gluten it naturally contains. So we eat something more nutritive along with our rice or potatoes—we add fat to our cabbage—we enrich our salad with vegetable oil—eat our cauliflowers with melted butter—and beat up potatoes and cabbage together into a nutritious kol-cannon.

Third—And thirdly, that in all natural varieties of vegetable food which are generally suitable for eating without cooking, a large per-centage of water is present. In preparing food in our kitchens we imitate this natural condition. Even in converting our wheaten flour into bread, we, as an important result aimed at, mix or unite it with a large proportion of water.

All the kinds of food by which the lives of masses of men are sustained being thus constituted, it is obvious that those vegetable substances which consist of one only of the constituents of wheaten bread, cannot be expected to prove permanently nutritious; and experience has proved this to be the case. The oils or fats alone do not sustain life, neither does starch or sugar alone. With both of these classes of substances, as we have seen, a certain proportion of gluten is associated in all our grains, fruits, and nutritive roots.

Hence arrow-root, which is only a variety of starch, cannot give strength without an admixture of gluten in some form or other. To condemn a prisoner to be fed on arrow-root alone, would be to put him to certain death by a lingering, torturing starvation. The same is true, to a less extent, of tapioca, and of most varieties of sago,* all of which consist of starch, with only a small and variable admixture of gluten. Even gluten, when given alone to dogs, has not kept them alive beyond a few weeks; so that no vegetable production, it may be said, and no kind of artificially pre-

* The pith of the sago palm, as it is made into bread by the natives of New Guinea, probably contains a sufficient proportion of gluten to sustain life; but this is in a great measure washed out in manufacturing the sago of commerce. Tapioca, as it is imported and consumed in this country, contains, I find, about three per cent. of gluten.

pared food, will support life, in which starch and gluten at least are not united. If they contain at the same time a certain proportion of fat, they will admit of more easy digestion, and of a more ready application in the stomach to the purposes of nutrition; and if they are either naturally permeated with a large quantity of water, or are transfused with it by artificial means, they will undergo a more complete and easy dissolution in the alimentary canal, and will produce the greatest possible effect in ministering to the wants of animal life.

It is interesting to observe how very generally adjustments of this kind have been made to the wants of animals, in the natural composition of the eatable parts of plants. But it is still more interesting to observe how experience alone has almost everywhere led men to a rude adjustment, in kind and quantity, of the forms of nutritive matter which are essential to the supply of their animal wants under the circumstances in which they are placed. And the absolute necessity of such adjustment is proved by all physiological history. For when, through force of circumstances, or through distorted taste, the natural instinct for such adjustment cannot be gratified, or is foolishly thwarted, the health is endangered, the constitution gradually altered, the temperament modified, life shortened, families extinguished, and whole races of men swept from the face of the earth. Such, looked at in their final effects, are the influences of the kind of food in which individuals indulge, or by which nations are supported.

CHAPTER VI.

THE BEEF WE COOK.

The fibrin and water of beef.—Composition of beef compared with that of wheaten bread and wheaten flour.—Striking differences.—Dried flesh compared with dried oat-cake.—More fat in domesticated animals and such as are fed for the butcher.—Composition of fish.—Richness of the salmon and the eel.—Less fat in fowls.—Eating butter with fish.—Composition of the egg.—Albumen or white; its properties and relations to gluten and fibrin.—Oil in the yolk, and in the dried egg.—Composition of milk.—Milk allied both to animal and vegetable forms of food.—Milk a model food.—Importance of a mixed food, containing much liquid.—Adjustment of the several ingredients of food in cooking.—Qualities of different kinds of cheese.—Composition of new and skimmed milk cheeses.—Comparison with milk.—Cheese as a digester.—Solvent power of decayed cheese.—Customary practices in cooking.—Qualities of different kinds of animal food.—Loss of beef and mutton in cooking.—Effects of heat upon meat.—Constituents of the juice of meat.—Kreatine.—Effects of salt upon meat.—Loss of nutritive value in salting.—How to boil meat and make meat soup.—Animal fats; their analogy to vegetable fats.—The solid fat of beef, mutton, and palm-oil.—Composition of human fat, goose fat, butter, and the oil of the egg.—The liquid part of animal fat.—Identity of animal and vegetable food as regards the mineral matters they respectively contain.

Beef and bread are the staples of English life; and as the study of wheaten bread in the preceding chapter gave us the key to the composition and nutritive qualities of all other vegetable substances, so an examination of beef will help us to a clear knowledge of all other kinds of animal food.

1°. Flesh. If a piece of fresh beef be dried in the hot sunshine, or in a basin, over boiling water, it will shrink, dry up, diminish in bulk, and lose so much water, that four pounds of fresh, newly-cut beef will leave only one pound of dried flesh.

Again, if we take a piece of lean beef and wash it in separate portions of clean water, its colour will gradually disappear. The blood it contains will be washed out, and a white mass of fibrous tissue will remain. If this be put into a bottle with alcohol or ether, a variable proportion of fat will be dissolved out of it, and the whole fibrous mass will now be dryer and more compact than before. Through this fibrous mass many minute vessels are scattered, but it chiefly consists of a substance to which chemists, from its fibrous appearance, give the name of *fibrin.*

The annexed woodcut (fig. 27) shows the structure of muscle, as seen under the microscope. The cross wrinkles represent the way in which the fibres contract in the living animal.

Fig. 27.

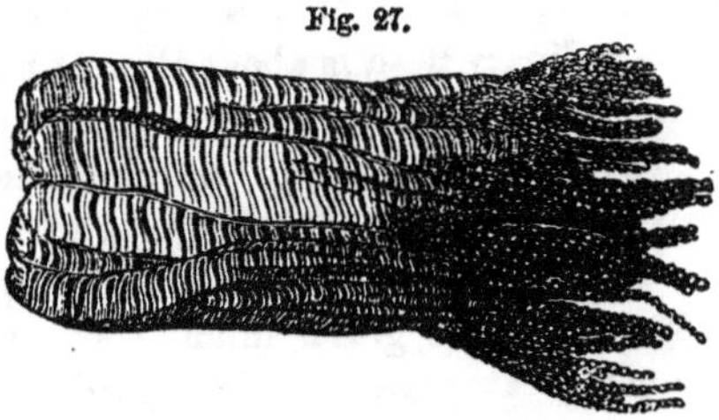

The fibres of lean muscle, showing how they are disposed or arranged,—the particles of which they are composed, and how they shrink or contract.

Of this fibrin the lean part of the muscles of all animals chiefly consists; it is therefore the principal constituent of animal flesh. It resembles the gluten of plants very closely in composition and properties—insomuch that, in a general comparison of animal with vegetable food, we may consider them for the present as absolutely identical.

Thus we have separated our beef—besides the small

quantity of blood and other matters washed out of it by the water—into three substances, water, fibrin, and fat. Its composition, as compared with that of wheaten bread and wheaten flour, it represented as follows:—

	Lean beef.	Wheaten bread.	Wheaten flour.
Water (and blood), . .	78	45	16
Fibrin or gluten, . .	19	6	10
Fat, ...	3	1	2
Starch, &c., . . .	....	48	72
	100	100	100

Lean beef, therefore, agrees with wheaten flour and bread, in containing water and fat—only in beef the water is as great as it is in the potato or the plantain. It agrees with them also in containing a substance (fibrin) which represents in the animal the gluten of the plant. The main difference between beef and bread are *first*, that the flesh does not contain a particle of starch, which is so large an ingredient in plants; and, *second*, that the proportion of fibrin in ordinary flesh is about three times as great as in ordinary wheaten bread. Or a pound of beef-steak is as nutritive as three pounds of wheaten bread, in so far as the nutritive value of food depends upon this one ingredient. In the dry matter of flesh, also, the proportion of fibrin is greater than that of gluten in any known vegetable food, and very much greater than in dried bread made from any of our cultivated grains.

This latter fact will become more apparent if we compare perfectly dry flesh with perfectly dry oat-cake—oatmeal being the richest of our common kinds of meal, both in gluten and in fat.

	Dried flesh.	Dried oat-cake.
Fibrin or gluten, ...	84	21
Fat, ...	7	7
Starch, ...	...	70
Blood and Salts, ...	9	2
	100	100

Here we have the two differences between the lean flesh of animals and the most nutritive of our grains presented in a very striking light. The animal food contains four times as much of what for the moment we may call gluten; but it is wholly deficient in the other main ingredient of vegetables—the starch—which in the dried oatmeal forms seven-tenths of the whole weight.

The flesh of wild animals is represented very nearly by the lean beef of which the composition is given above. Wild animals generally contain little fat. But it is not so with our domesticated animals, and especially such as are reared for food. They all contain much fat, either collected by itself in various parts of the body (the suet or tallow), or intermingled with the muscular fibre, as in the highly-prized marbled beef in which the English epicure delights. In the boiling-houses at Port Philip, a small merino sheep of 55 lb. weight gives 20 lb. of tallow, which is nearly two-fifths of the whole. In heavier sheep the proportion of fat increases, four-fifths of all the weight above 55 lb. being tallow. In beef and mutton, such as is met with in our markets, from a third to a fourth of the whole dead weight generally consists of fat.

Supposing that, as it comes to the table, one-fourth of the weight of the butcher-meat we consume consists of fat, then the nutriment contained in 100 lb. of it, made quite dry, will be represented by—

Fibrin,	63 lb.
Fat,	30 „
Salts and blood,	7 „
	100

This fat to a certain extent represents and replaces the starch of vegetable food.

Fowls contain less fat than butcher-meat; though, when crammed and fed upon food rich in fat, the capon and the ortolan, and the diseased livers of the goose, become as rich as the fattest beef or mutton.

The composition of other kinds of flesh which we eat as food is much the same as that of beef. Veal and venison contain less fat, while pork contains more. Each variety also possesses a peculiar flavour and a faint odour, which is characteristic of the species, and sometimes of the variety of the animal. In some cases, as with our mountain mutton, this peculiar flavour is a high recommendation; in others, as with the sheep of the Low Countries, and with the goat, it renders them to many altogether unpalatable.

2°. FISH in general is less rich in fat than the flesh meat in our markets, and consequently contains more fibrin. Some of our common varieties of fish, when perfectly dried, consist of—

	Fibrin.	Fat, &c.
Skate, ...	97	3
Haddock, ...	92	8
Herring, ...	92	8
Salmon, ...	78	22
Eel, ...	44	56

These numbers, of course, are liable to variation—the herring especially being very much fatter at some seasons and on some coasts than on others. We see, however, that salmon is justly considered a *rich* fish, since it contains three times as much fat as the haddock. The epicure has also a substantial reason for his attachment to the eel, since it contains a considerably greater weight of fat than it does of muscular fibre.

It appears, therefore—

First, That the dried flesh of all the animals that we most usually consume for food, consists essentially of fibrin.

Second, That the proportion of fat is variable, and that

those varieties of animal food are most esteemed for human food in which a considerable proportion of fat is present. Hence,

Third, Where the proportion of fat is naturally small, we endeavour to increase it by art; as in feeding the capon. Or we eat along with those varieties in which it is small some other food richer in fat. Thus, we eat bacon with veal, with liver, and with fowl; or we capon the latter, and thus increase its natural fat. We use melted butter with our white fish, or we fry them with fat; while the herring, the salmon and the eel, are usually both dressed and eaten in their own oil. If the reader will take the trouble of consulting any popular cookery-book, he will find that sausage, and other rich mixed meats, are made in general with one part of fat and two of lean—the proportion in which they exist in a piece of good marbled beef! Art thus unconsciously again imitating nature.

3°. THE EGG.—Akin to flesh and fish is another form of animal food—the egg. The egg of the domestic hen is that which is most commonly known, and most extensively used as food. It consists of three principal parts—the shell, the white, and the yolk. The shell is composed of carbonate of lime or hard chalk, and it is intended chiefly as a protection to the inner part. It is penetrated, however, by numerous minute holes or pores, through which the air is capable of passing, and by means of which it is conveyed to the young bird during the process of hatching.* It forms rather more than a tenth part of the weight of the egg, the white forms six-tenths, and the yolk three-tenths. A com-

* Through these pores, also, the air enters, by the agency of which eggs, when kept, soon become rotten. If those pores are filled up by rubbing the new-laid egg over with fat, or in any similar way, it will keep fresh for an indefinite period. It is then very nearly in the condition of the hermetically sealed meats now prepared for use in long voyages.

mon-sized hen's egg weighs about a thousand grains, and consists, therefore, of about—

White, ...	600 grains.
Yolk, ...	300 "
Shell, ...	100 "
	1000 grains.

The white of the egg is so called, because, when heated, it coagulates into a white solid substance, which is insoluble in water, and almost free from taste. It is known to chemists by the name of *albumen.** Though different in appearance and in sensible properties from fibrin and gluten, it has a very close chemical relation to these substances, and serves nearly the same purpose in the feeding of animals. We may for the present, therefore, consider all the three—gluten, fibrin, and albumen—as, in a nutritive sense, absolutely identical.

The yolk is of a yellow colour. It consists, in part, of a variety of albumen, and, therefore, like the white, coagulates, though in a less degree, when the egg is heated. But if the dry hard yolk be crushed, and digested in alcohol, or in ether, it becomes colourless, while the spirit extracts, and dissolves a bright yellow oil. This oil forms about two-thirds of the weight of the yolk, in its perfectly dry state. Thus the yolk, like flesh and fish, consists of fat intermixed with a substance which has a close resemblance to the gluten of plants.

The egg contains, besides, a large per-centage of water, amounting, as in fresh butcher-meat, to nearly three-fourths of its whole weight. Thus the egg, when deprived of its shell, consists, in the natural and in the dried states, respectively of—

* See note, p. 81.

	Natural state.		Dried at the heat of boiling water.
	Whole egg.	Per cent.	Per cent.
Water, ...	666	74	—
Albumen, ...	127	14	54⅔
Fat, ...	94	10½	40
Ash (when burned), . .	13	1½	5⅓
	900	100	100

It contains also a trace of milk-sugar.

The egg, therefore, as a whole, is richer in fat than fat beef. It is equalled, in this respect, among common kinds of food, only by pork and by eels. It is of interest to remark, however, that the white of the egg is entirely free from fat, and that albumen is a very constipating variety of animal food, so that it requires much fat to be eaten along with it, when consumed in any quantity, in order that this quality may be counteracted. It is, no doubt, because experience has long ago proved this in the stomachs of the people, that "eggs and bacon" have been a popular dish among Gentile nations from time immemorial.

4°. MILK.—Another nutritious form of animal food is the well-known fluid milk. This, as we should expect, contains more water than beef or the egg; yet, contrary to what we might expect, not more than the turnip, and much less than the melon.

Milk, by one well-known process, yields butter or fat, and by another curd or cheese. The curd, to which chemists give the name of *casein*, from its forming cheese, resembles the gluten, fibrin, and albumen, of which we have already spoken, and is classed along with them as a nutritive substance. It possesses also, weight for weight, about the same value, when used as food, and, like albumen, is distinguished, when eaten alone, for a remarkably constipating property.

When the whey of milk, from which the curd and butter have been completely separated, is evaporated to dryness, a colourless sweet substance is obtained, which is known by the name of sugar of milk. When dried and burned in the air, milk also leaves behind a quantity of ash. These several ingredients exist in cow's milk, in the natural and in the dried states, in the following average proportions :—

	Natural state.	Evaporated to dryness.
Water,	87	—
Curd, or casein,	4½	34¾
Butter, or fat,	3	23¼
Sugar (of *milk*),	4¾	37
Ash (*nearly*),	¾	4½
	100	100

Thus milk appears to partake of the nature of both animal and vegetable food. It contains a large proportion of curd and butter, which represent the fibrin and fat of beef, and, at the same time, a large proportion of sugar, which represents the starch of wheaten bread.

Human milk very closely resembles the milk of the cow. Its average composition is as follows :—

Water,	88.91	or 89
Curd or casein,	3.92	— 4
Butter or fat,	2.67	— 2⅔
Sugar of milk,	4.36	— 4⅓
Salts or ash,	0.14	— 1-7
	100	100 1-7

The principal difference is in the proportion of saline matter, which in human milk is only one-third of that of cow's milk.*

* The milk of women from fifteen to twenty years of age contains more solid constituents than that of women between thirty or forty. Women with dark hair also give a richer milk than women with light hair. In *acute* diseases the sugar decreases one-fourth and the curd increases one-fourth, while in *chronic* affections the butter increases one-fourth, and the casein slightly diminishes. In both classes of diseasos the proportion of saline matter increases.

Now, as the natural food of the young mammalian animal of every species is the milk of its mother, that milk may be looked upon as a kind of model food for the species to which the animal belongs. Woman's milk, therefore, is the type of human food, and after its form and composition all other kinds of food should be adjusted, especially in the case of persons whose condition approaches to that of the child. Hence it seems reasonable to infer—

First, That our food ought to contain a due admixture of vegetable and animal food substances, in which the proportions of the three most important constituents, fat, starch or sugar, and fibrin or gluten, are properly adjusted.

Second, That the food, if not naturally liquid, should be intimately mixed with a large quantity of liquid before it is introduced into the stomach. This lesson we have already learned from the study of various natural forms of vegetable food.

The attainment of these two ends, in such a way as at the same time to please the eye and the palate, guides, for the most part, the operations of the cook in his kitchen. They ought always to guide the operations of those who wish to prepare what it will be wholesome for the majority of men to eat.

5°. CHEESE.—The manufacture of cheese of different varieties, and the qualities which these varieties severally possess, are illustrations of the importance of a mixed food.

Cheese is eaten for two very different purposes—either as a part of the regular food, for the general sustenance of the body, or as a kind of condiment, taken in small quantity along with or after the usual fare as is common at dinner-tables.

In the making of cheese many different varieties are obtained, according as the proportion of cream is increased or diminished. When it is made from cream alone, what is called a *cream cheese* is obtained, which must be used when comparatively fresh, as it soon becomes rancid. When the cream of the previous night's milking is added to the new milk of the morning a very rich cheese is made, like our English Stilton; when good new milk only is employed, rich cheeses like the Cheddar are obtained; when an eighth or tenth of the cream is removed, highly esteemed cheeses, like the large-sized (120 lb.) Cheshires are made, which will not hold together if all the cream be left in. There seems, at first sight, to be no connection between the application of bones to the Cheshire farmer's poor grass-land and the unexpected crumbling of the Cheshire dairymaid's cheese. Yet the connection is plain enough. The bones bring up richer grass; this gives richer milk; and this, treated in the old way, a fatter and therefore more crumbly cheese. When the skimmed milk of the evening is added to the new milk of the morning, the mixed milk yields cheeses like the single Glo'ster. If the cream be once removed from the whole of the milk, it yields common skimmed-milk cheese; if it be twice creamed, it gives cheeses like some of the poorer sorts made in Friesland; and if skimmed for three or four days in succession, it yields the hard horny cheeses of Suffolk, locally known by the name of *Suffolk bank*, which often requires an axe to cut it, and which is so hard "that pigs grunt at it, dogs bark at it, but neither of them dare bite it."

Now, in the making of cheese, the milk is first curdled—sometimes by the use of vinegar, but generally by means of rennet. The curd is then separated from the whey, in which the sugar of milk remains dissolved; after this it is carefully pressed and dried. Were there no cream taken off the milk, therefore, the cheese as a food would differ from the milk chiefly in containing little or no sugar. But when more or less of the cream is removed from the milk employed, the cheese becomes further removed from milk in its composition, and less fitted, therefore, to serve alone as a nutritious animal diet. The following numbers represent the composition of a rich Cheddar cheese when two years old, and of a common one-year-old skimmed-milk cheese made in Lanarkshire.

	Cheddar.	Skim milk.
Water, ...	36	44
Curd, ...	29	45
Fat, ...	30½	6
Ash, ...	4½	5
	100	100

Both contain a very considerable proportion of water, and therefore in this respect they are not unsuited for immediate consumption as food. But while the fat in one amounts to nearly one-third of the whole weight, in the other it only reaches to six per cent.

But we shall have a clearer idea of the value of these varieties of cheese for a general diet, by comparing their composition in a dried state with those of milk, beef, and eggs, also in a dried state. This is seen in the following table:—

	Milk.	Cheese.		Beef.	Eggs.
		Cheddar.	Skim milk.		
Casein (curd), . .	35	45	80	89*	55
Fat (butter), . .	24	48	11	7	40
Sugar, ...	37	...	....	....	....
Mineral matter, . .	4	7	9	4	5
	100	100	100	100	100

* This number is something larger than that given in page 108. This is because the weight of the blood (five per cent.), which consists chiefly of fibrin and albumen, is here added to that of the fibrin of the beef in which it is contained. The reader will not forget that casein, fibrin, and albumen are all nearly identical with each other and with the gluten of plants.

We see from this table that both cheeses are free from sugar. Either of them, therefore, must be eaten with a quantity of vegetable food which may supply the starch or sugar required to make it equal to milk as a general nourishment. Again, the Cheddar cheese contains more fat even than the egg. It is too rich, therefore, to be used as an everyday diet by the generality of stomachs. It is partly for this, and partly for the previous reason, that "cheese and bread" are almost invariably eaten together.

Then, in the skim-milk cheese, we have only eleven of fat mixed with eighty of the very constipating curd. Experience has shown this to be far too little, and therefore butter or fat bacon, as well as bread, must be consumed along with these poorer cheeses, when much of them is intended to be eaten; or they must be cooked, in made dishes, along with some other variety of fat.

It is with a view to similar adjustments in the proportions of the several necessary ingredients of a nourishing food, that we mix eggs with sago, tapioca, and rice in our puddings, shred the oily yolk into our salad, boil rice with milk, and eat rich cheese with our maccaroni.

But cheese is often eaten also as a relish or condiment, only in small quantities at a time. It is chiefly the older and stronger-tasted varieties that are so used. They are generally very wholesome and digestible when taken in this way. As a *digester*, as some not inappropriately call it, cheese—that which is decayed and mouldy being preferred by connoisseurs—is often eaten after dinner. The action which experience seems to have proved it to possess, in aiding the digestion of what has previously been eaten, is both curious and interesting, and has had some light thrown upon it by recent chemical research.

When the curd of milk is exposed to the air in a moist state for a few days at a moderate temperature, it begins gradually to decay, to emit a disagreeable odour, and to ferment. When in this state, it possesses the property, in certain circumstances, of inducing a species of chemical change and fermentation in other moist substances with which it is mixed, or is brought into contact. It acts after the same manner as sour leaven does when mixed with sweet dough.

Now, old and partially decayed cheese acts in a similar way when introduced into the stomach. It causes chemical changes gradually to commence among the particles of the food which has previously been eaten, and thus facilitates the dissolution which necessarily precedes digestion. It is only some kinds of cheese, however, which will effect this purpose. Those are generally considered the best in which some kind of cheese mould has established itself.* Hence the mere eating of a morsel of cheese after dinner does not necessarily promote digestion. If too new or of improper quality, it will only add to the quantity of food with which the stomach is probably already overloaded, and will have to await its turn for digestion by the ordinary processes.

We have seen that it is one of the special advantages possessed by the varieties of flour obtained from wheat and rye, that in the hands of the baker they form light and spongy bread. This is owing, as I have explained, to a peculiarly tenacious property which is possessed by the kinds of gluten contained in these two species of grain. But the same property is possessed to some extent by the white of the egg. It has a glairy consistence, which enables it, when mixed up with moistened flour, arrow-root, sago, &c., to retain the globules of air or of steam which are produced with-

* It is an interesting circumstance that such kinds of cheese mould, and the flavour and digestive quality which accompany them, may be propagated even in newer cheeses by inoculation,—removing a bit of the new, that is from the interior, and putting in a bit of the old in its place.

in it by fermentation or by heat. Thus, like the gluten of wheat, it enables the mixed materials to swell up into a porous mass. Hence the lightness which the white of egg gives to puddings, to cakes, and even to wheaten bread. In a less degree, a similar quality resides in the curd of milk, and hence one cause of the improvement in the appearance of bread which has been wholly or in part prepared with milk.

Before leaving this part of the subject, it may be useful to exhibit in a tabular form the composition of dried beef, eggs, and milk, compared with that of dried wheaten flour and dried oatmeal.

	Beef.	Eggs.	Milk.	Fine wheaten flour.	Oatmeal.
Fibrin, casein, albumon, or gluten,	89	55	35	12	21
Fat, ...	7	40	24	2½	7
Starch or sugar, . . .	—	—	37	83½	70
Ash or mineral matter, . .	4	5	4	2	2
	100	100	100	100	100

From this table many interesting comparative deductions may be drawn.

6°. Cooking flesh meat.—In cooking animal food, plain boiling, roasting, and baking, are in most general favour in our islands. During these operations, fresh beef and mutton, when moderately fat, lose on an average about—

	In boiling.	In baking.	In roasting.
4 lb. of beef loso,	1 lb.	1 lb. 3 oz.	1 lb. 5 oz.
4 lb. of mutton lose,	14 oz.	1 lb. 4 oz.	1 lb. 6 oz.

The greater loss in baking and roasting arises chiefly from the greater quantity of water which is evaporated, and of fat which is melted out during these two methods of cooking. Two circumstances, however, to which it has not hitherto been necessary to advert, have much influence upon the successful result of these and some other modes of cooking.

If we put moist flesh meat into a press and squeeze it, a red liquid will flow out. This is water coloured by blood, and holding various saline and other substances in solution. Or if, after being cut very thin, or chopped very fine, the flesh be put into a limited quantity of clean water, the juice of the meat will be gradually extracted, and by subsequent pressure will be more completely removed from it than when pressure is applied to it in the natural state, and without any such mincing and steeping. The removal of these juices leaves the beef or mutton nearly tasteless.

When the juice of the meat extracted in either way is heated nearly to boiling, it thickens or becomes muddy, and flakes of whitish matter separate, which resemble boiled white of egg. They are, in fact, white of egg or albumen, and they show that the juice of flesh contains a certain quantity of this substance in the same liquid and soluble state in which it exists in the unboiled egg. Now, the presence of this albumen in the juice of butcher meat is of much importance in connection with the skilful preparation of it for the table.

The first effect of the application of a quick heat to a piece of fresh meat is to cause the fibres to contract, to squeeze out a little of the juice, and to a certain extent to close up the pores so as to prevent the escape of the remainder. The second is to coagulate the albumen contained in the juice, and thus effectually and completely to plug up the pores, and to retain within the meat the whole of the internal juice. Thereafter, the cooking goes on through the agency of the natural moisture of the flesh. Converted into vapour by the heat, a kind of steaming takes place within the piece of meat, so that whether in the oven, on the spit, or in the midst of boiling water, it is in reality, when skilfully done, cooked by its own steam.

A well-cooked piece of meat should be full of its own juice or natural gravy. In roasting, therefore, it should be exposed to a quick fire, that the external surface may be made to contract at once, and the albumen to coagulate, before the juice has had time to escape from within. And so in boiling. When a piece of beef or mutton is plunged into boiling water, the outer part contracts, the albumen, which is near the surface, coagulates, and the internal juice is prevented either from escaping into the water by which it is surrounded, or from being diluted and weakened by the admission of water among it. When cut up, therefore, the meat yields much gravy and is rich in flavour. Hence a beef-steak or a mutton-chop is done quickly, and over a quick fire, that the natural juices may be retained.

On the other hand, if the meat be exposed to a slow fire, its pores remain open, the juice continues to flow from within as it is dried from the surface, and the flesh pines and becomes dry, hard, and unsavoury. Or if it be put into cold or tepid water, which is afterwards gradually brought to a boil, much of the albumen is extracted before it coagulates, the natural juices for the most part flow out, and the meat is served in a nearly tasteless state. Hence, to prepare good boiled meat, it should be put at once into water already brought to a boil. But to make beef-tea, mutton broth, or other meat soups, the flesh should be put into the cold water, and this afterwards very slowly warmed, and finally boiled. The advantage derived from *simmering*, a term not unfrequent in cookery books, depends very much upon the effects of slow boiling as above explained.

7°. Beef-tea.—It has lately been recommended to make beef-tea by simply chopping the meat small, pouring upon it its own weight, or any other desired quantity of cold water, and bringing it quickly to a boil. This process extracts all the natural juices and gives a most agreeable and savoury tea, which holds in solution about one-eighth part of the solid substance of the beef. But it has been stated, as a recommendation of this process, *first*, that the tea, obtained contains *all* the nutritive qualities of the meat, which is said to be no longer of any value, and, *second*, that it is as nutritious as if the meat were boiled long enough to give a tea which should stiffen to a jelly when cold.

But this statement is incorrect, and is made only in consequence of two very opposite things being confounded. The juice of the meat contains a small proportion of a substance called *kreatin*, which is rich in nitrogen, has a certain chemical relation to the peculiar principle of tea and coffee (*thein*)—of which I shall speak in a subsequent chapter—and exercises, as I believe, a special tonic and exhilarating influence upon the system, independent of any directly nutritive quality it may possess. This substance, with all the soluble salts of the flesh, the beef-tea made after the above process contains, and the residual fleshy fibre is tasteless, and will not alone support animal life for any length of time. But eaten along with the tea thus made, or with what the tea contains, or made into savoury meat by the addition of ordinary gravy, it will sustain and strengthen the body, as all experience proves. The meat tea also will be more nutritious, in the ordinary sense, the more of the jelly-forming substance of the meat it holds in solution. It will bear, in fact, to the thinner and more quickly made beef-tea, a similar relation to that which cocoa bears to the infusion of China tea.* Both of these last named beverages contain a peculiar principle rich in nitrogen, which exercises

* See THE BEVERAGES WE INFUSE.

a special influence on the activity of the brain; but the cocoa is rich besides in the substances which form our ordinary nourishment. And as, in consequence of this difference, cocoa is not so well suited as tea or coffee to the digestive powers of some constitutions, so it probably is with the meat teas or decoctions prepared by the two processes referred to. The correct values, both relative and absolute, of the meat teas made after the two methods, as well as of the undissolved residue of the meat, are therefore easily seen and understood.

8°. SALTING of Meat.—The application of salt to fresh meat has very much of the same effect as the application of a quick heat. It causes the fibres to contract, the meat to lesson in bulk and the juice to flow out from its pores. Hence the reason why dry salt strewed upon fresh lean meat gradually dissolves into a fluid brine. The effect of the salt, if a large quantity be applied, penetrates deep, so that as much as one-third of the juice of the meat is often forced out by the contraction of the fibres. The effect of this upon the meat is twofold. It diminishes the natural flavour, by removing a large proportion of the peculiar substances contained in the juice, and adding pure salt in their stead. At the same time it closes up the pores of the meat, and prevents the entrance of atmospheric air, thus diminishing the liability to decay.

The preservation of flesh meat by salting, depends, therefore, upon the separation of water, upon the exclusion of air, upon the saturation with salt of the juice which remains in the meat, and upon the formation of a weak compound of the flesh with common salt, which does not readily undergo decay. But this preservation is attended by a diminution in its nutritive qualities, for the juice which flows out contains albumen (white of egg), kreatin, phosphoric acid, and potash. These substances are precisely the same as are more fully extracted by water, in the method of making savoury beef-tea, already described, and in proportion as they are extracted they diminish the nutritive properties of the meat. Hence one reason why long feeding on salt meat affects the health, and why vegetable and other substances, which are capable of supplying what the meat had lost, are found to be the best means of restoring it.

As a whole, flesh meat is eminently nutritious, because it contains *all* the materials which are necessary to build up our own flesh; but remove from it a portion of these materials, and the remainder becomes more or less useless,—as bricks and stone become useless to the builder if we refuse him the requisite quantity of mortar.

9°. The FAT of Animal and Vegetable Substances.—We have seen that, as a whole, there is much analogy between the bread and the beef,—the vegetable and the animal forms of food on which we live. Between the gluten of the one and the fibrin of the other, we have also found a very close similarity, and that in the animal economy they are both fitted and intended to serve the same main purpose. If we compare the fatty portions of both, we find new resemblances.

Most of the varieties of fat yielded by our common European vegetables are fluid and oily at ordinary temperatures. Such is the case with the fat extracted from wheat, from oats, from Indian corn, from linseed, from the olive, the poppy, the walnut, &c. The fat of the oil palm, however, commonly known by the name of palm-oil, and some other vegetable fats or butters, are solid in the natural state, and at ordinary temperatures. And even the oily fats, olive oil for example, when exposed to a low temperature, congeal or freeze to a certain extent, and allow of the separation of a solid fat in greater or less proportion. On the other hand, those which are solid yield to pressure a quantity of a liquid fatty oil. So that in reality all vegetable fats consist of two fatty substances, one of which is solid, and the other liquid, at ordinary temperatures.

Now, the same is the case with the animal fats—with those of beef and mutton for example, with the butter of milk, and with the oil contained in the yolk of the egg. All consist of a solid and a liquid fat, and in this fact we see a new analogy between our vegetable and our animal food.

But a still further and more intimate analogy exists between the solid portions of the fatty substances of the animal and vegetable kingdoms. When the solid fat of palm-oil is properly purified it is found to consist of a solid, beautifully white, peculiar fatty body, to which the name of *palmitine* has been given. On the other hand, when beef and mutton fats are pressed from the oil they contain, and then purified, the most abundant substance obtained is a peculiar fat which is known by the name of *stearine*. The remainder consists principally of palmitine.

Now, of these two fatty bodies the solid fat of all our domestic animals almost entirely consists. In beef and mutton fats the stearine is the more abundant. In human fat, in that of the goose, and in that of butter, the stearine and palmitine are in nearly equal proportions. It is the same with vegetable fats. They consist of these two varieties in different proportions. In some the solid part consists chiefly of stearine; in others, as in olive-oil, the stearine and palmitine are nearly equal in quantity; while in others again, as in palm-oil, the palmitine is the principal ingredient. Thus, as there is a kind of identity in nutritive quality and value among the compounds represented respectively by gluten in plants and by fibrin in animals, so there is an absolute identity of substance—as regards their solid part at least—among the fatty compounds which are met with in the eatable productions of both kingdoms.

The liquid portions of the fats of animals and vegetables, though generally regarded as being also for the most part identical, are not yet so well understood as their solid portions. It is a fact of practical interest, however, that they become rancid by exposure to the air sooner than the solid fats do. Hence hard butter keeps sweet longer than soft butter does. Hence, also, fat meat keeps longer, when salted, if the fat be hard. And hence the reason why, in finishing off fat animals for the butcher, especially if they are to be salted, it is usual to give dry food for some time before killing, that the fat may be hardened and the flesh made firm.

In another matter of detail I might show how, in still more minute matters, animal and vegetable kinds of food are nearly identical. When the parts of plants are burned in the open air they disappear for the most part, as I have already shown,* and leave only a small proportion of ash behind. This ash consists of a mixture of various substances, spoken of as their mineral, earthy, saline, or inorganic constituents.

The same takes place when the parts of animals are burned; and the mixture of mineral matters obtained consists, in either case, of the same substances, only differing more or less in their relative proportions. The same things occur in the ash of bread as are found in the ash of beef. In whatever degree, therefore, the nutritive properties of our food depend upon the kind of mineral matter it contains, it is almost a matter of indifference whether we live upon an animal or a vegetable diet.

But to this interesting point I shall have occasion to return in a subsequent chapter.

* See THE PLANT WE REAR, p. 68.

CHAPTER VII.

THE BEVERAGES WE INFUSE.

THE TEAS.

Artificial drinks nearly all vegetable infusions, with or without subsequent chemical changes.—Tea, extensive use of.—The tea-plant; how its leaves are gathered.—The aroma produced by the roasting.—Mode of preparing green and black tea from the same leaves.—Principal varieties of green and black tea.—Differences in fragrance and flavour.—Ancient use of tea in China and the adjoining countries.—Introduction into Europe.—Total amount of tea produced.—Consumption in the United Kingdom.—Sensible effects of tea.—Active chemical ingredients in tea.—The volatile oil, its action.—The theine, its composition.—Occurs in coffee, in maté, and in guarana.—Its effect in retarding the waste of the tissues.—Why tea is a favourite vith the poor.—The tannin, its properties and effects.—The gluten.—Tea-leaves and beans compared in nutritive quality.—Tartar mode of using tea.—Eating the exhausted leaves.—Tea varies in composition.—Proportion extracted by water varies.—How tea is coloured or dyed green in China.—Lie tea.—Maté or Paraguay tea.—Its ancient use in South America.—The *Ilex paraguayensis* or maté tree, where it grows, and how its leaves are collected.—Gongonha of Brazil, a variety of maté.—Frequent use of maté, and its effects.—Composition of the leaf.—The volatile oil, the theine, the tannic acid, and the gluten.—Coffee-tea made from the leaf of the coffee tree.—Use of this tea in the Eastern Archipelago.—Effects observed from its use in Sumatra.—Contains the same active ingredients as the leaves of the tea trees.—Labrador tea used in North America.—Abyssinian tea or chaat.—Tasmanian teas.—Faham tea.—Substitutes for Chinese tea and for maté.

THE two most important natural liquids, water and milk, have already been treated of; various artificial drinks, however, are prepared both in civilised and in semi-barbarous countries, and are in daily use among vast multitudes of men;—such as tea, coffee, and cocoa, beer, wine, and ardent spirits—the preparation and effects of each of which are connected with interesting chemical considerations.

These drinks agree in being all prepared from or by means of substances of vegetable origin, and in being generally classed among the luxuries, rather than the necessaries of life.

The mode in which they are prepared, however, naturally divides these drinks into two classes. Tea, coffee, and cocoa are roasted and prepared before they are infused in water, and the infusion is then drunk without further chemical treatment. These are simple *infused* beverages. Beer, wine, and ardent spirits are prepared from infusions which, after being made, are subjected to important chemical operations. Among these operations is the process of fermentation, and hence they are properly distinguished as *fermented* liquors.

I shall consider these two classes of drinks, therefore, separately, and in the order in which I have mentioned them.

The infused beverages are drunk hot, fermented drinks are usually taken cold. The love of such warm drinks prevails almost universally. In frozen Labrador and snowy Russia, the climate might account for this predilection, but the craving is really deeper seated. The practice prevails equally in tropical and in arctic regions. In Central America the Indian of native blood, and the Creole of mixed European race, indulge alike in their ancient chocolate. In Southern America the tea of Paraguay is an almost universal beverage. The native North American tribes have their Apallachian tea, their Oswego tea, their Labrador tea, and many others. From Florida to Georgia, in the United States, and over all the West India islands, the naturalised European races sip their favourite coffee; while over the Northern States of the Union, and in the British provinces, the tea of China is in constant and daily use.

All Europe, too, has chosen its prevailing beverage. Spain and Italy delight in chocolate; France and Germany, and Sweden and Turkey, in coffee; Russia, Holland, and England in tea,—while poor Ireland makes its warm drink of the husks of the cocoa, the refuse of the chocolate mills of Italy and Spain.

All Asia feels the same want, and in different ways has long gratified it. Coffee, indigenous in Arabia or the adjoining countries, has followed the banner of the Prophet, wherever in Asia or Africa his false faith has triumphed. Tea, a native of China, has spread spontaneously over the hill country of the Himalayas, the table-lands of Tartary and Thibet, and the plains of Siberia—has climbed the Altais, overspread all Russia, and is equally despotic in Moscow as in St. Petersburg. In Sumatra, the coffee-leaf yields the favourite tea of the dark-skinned population, while Central Africa boasts of the Abyssinian *chaat* as the indigenous warm drink of its Ethiopian peoples. Everywhere unintoxicating and non-narcotic beverages are in general use,—among tribes of every colour, beneath every sun, and in every condition of life. The custom, therefore, must meet some universal want of our poor human nature.

The beverages we infuse naturally arrange themselves into three classes. First, the *teas* or infusions of leaves. Second, the *coffees* or infusions of seeds. And third, the *cocoas*, which are more properly soups or gruels than simple infusions, as they are made by diffusing, through boiling water, the entire seeds of certain plants previously ground into a paste.

I. The Teas.—Of teas there are many varieties in use in different parts of the world; but China tea, Paraguay tea or maté, and perhaps coffee-tea, are the most extensively consumed as national beverages. There are some others in constant though less general employment, to which it will be necessary somewhat briefly to advert.

1°. CHINA TEA is not only the most important of these beverages to the British and other English-speaking peoples, but it forms the daily drink of a larger number of men than all the others put together. Among the three hundred millions of China, and among the inhabitants of Japan, Thibet, and Nepaul, it is an article of consumption with all classes three or four times a-day. In Asiatic Russia also, in a large portion of Europe, in North America, and in Australasia, it is in, or is coming into, almost equally extensive use. It is consumed at the present moment by probably not less than five hundred millions of men, or one-half of the whole human race.

Fig. 28.

Thea bohea—the Bohea Tea-plant.
Scale, 1 inch to 5 feet.
Scale for leaf, 1 inch to 2 inches.

The tea-plant (*Thea sinensis*) has much resemblance to the Camellia Japonica. There are several varieties of it, distinguished by some botanists as the *Thea viridis*, *T. bohea*, and *T. strieta*, but all are now recognised as belonging to one single species, somewhat altered in habit and appearance by cultivation, climate, and soil. The two most marked varieties are represented by the annexed woodcuts. The smaller (fig. 28) is the *Thea bohea*, which produces the inferior green and black teas which are made about Canton. The larger (fig. 29) is the *Thea viridis*, the more northern variety, from which are made all the fine

Fig. 29.

Thea viridis—the common Tea-plant.
Scale, 1 inch to 5 feet.
Scale for leaf, 1 inch to 2 inches.

green teas in the great Hwuy-chow and the adjoining provinces. The plant is believed to be a native of China, and grows wild still among the hills both of that country and of Japan. It thrives best in the cooler parts of the tropical zone, but grows in the temperate zone even as far north as the 40th degree of north latitude. The districts of China which supply the greater portion of the teas exported to Europe and America lie between the 25th and the 31st degrees of north latitude, and the best districts are those between 27° and 31°.—(FORTUNE.)

The tea-plants are raised from seed which, to secure germination, is kept over winter in moist earth, and sown in March. When a year old, the young bushes are planted out, and then by cropping the main shoot for the first year they are kept down to a height of about 3 feet, and made to grow bushy. Being placed in rows 3 or 4 feet apart, they have some resemblance to a garden of gooseberry bushes. The cropping of the leaves begins in the fourth and fifth years, and is seldom continued beyond the tenth or twelfth, when the bushes are dug up and renewed. The plant thrives best on dry sunny slopes, where occasional showers fall and springs appear, and where an open, somewhat stony but rich soil prevents the water from lingering about its roots. The season for gathering varies in different districts, but the principal leaf-harvest ends in May or June. The leaves are plucked by the hand, and chiefly by women. They are generally gathered at three successive seasons. The youngest and earliest leaves are the most tender and delicate, and give the highest flavoured tea. The second and third gatherings are more bitter and woody, and yield less soluble matter to water. The refuse and decayed leaves and twigs are pressed into moulds and sold under the name of brick tea. These bricks are often made harder by mixing the leaves with the serum of sheep and ox blood. This inferior variety is chiefly consumed in northern China and Thibet.

The first in order, and not the least interesting point, in the chemical history of the tea we use, is the mode in which it is prepared for the market. The leaves when freshly plucked have neither a decidedly astringent, an aromatic, nor a bitter taste. They possess nothing, in fact, either of the odour or flavour of the dried leaves. The pleasant taste and delightful natural scent for which they are afterwards so highly prized, are all developed by the roasting which they undergo in the process of drying. The details of this process have lately been made known to us through the investigations of Mr. Fortune.

Another interesting chemical fact is, that different qualities of tea are prepared from the same leaves, according to the way in which they are treated in the drying. This we should to a certain extent expect; but the inquiries of Mr. Fortune have shown that samples so very unlike as the green and black teas usually are may be prepared at will from the same leaves, gathered at the same time and under the same circumstances. The mode of drying and roasting the leaves generally, and the specific processes by which the green and the black teas are severally obtained, have been minutely described by Mr. Fortune;* and from his description we learn—

* His description is as follows:—

For Green Tea.—When the leaves are brought in from the plantations they are spread out thinly on flat bamboo trays, in order to dry off any superfluous moisture. They remain for a very short time exposed in this manner, generally from one to two hours; this, however, depends much upon the state of the weather.

In the mean time time the roasting-pans have been heated with a brisk wood-fire. A portion of leaves is now thrown into each pan, and rapidly moved about and shaken up with both hands. They are immediately affected by the heat, begin to make a crackling noise, and become quite moist and flaccid, while at the same time they give out a considerable portion of vapour. They remain in this state for four or five minutes, and are then drawn quickly out and placed upon the rolling-table, and rolled with the hands.

Having been thrown again into the pan, a slow and steady charcoal fire is kept up,

First, That, in the process of drying, the leaves are roasted and scorched in such a way as necessarily to bring

and the leaves are kept in rapid motion by the hands of workmen. Sometimes they are thrown upon the rattan-table, and rolled a second time. In about an hour, or an hour and a half, the leaves are well dried, and their colour has become *fixed*,—that is, there is no longer any danger of their becoming black. They are of a dullish green colour, but become brighter afterwards.

The most particular part of the operation has now been finished, and the tea may be put aside until a larger quantity has been made. The second part of the process consists in winnowing and passing the tea through sieves of different sizes, in order to to get rid of the dust and other impurities, and to divide the tea into the different kinds known as twankay, hyson-skin, hyson, young hyson, gunpowder, &c. During this process it is re-fired—the coarse kinds once, and the finer sorts three or four times. By this time the colour has come out more fully, and the leaves of the finer kinds are of a dull bluish green.

For Black Tea.—When the leaves are brought in from the plantations they are spread out upon large bamboo mats or trays, *and are allowed to lie in this state for a considerable time.* If they are brought in at night they lie until next morning.

The leaves are next gathered up by the workmen with both hands, thrown into the air, and allowed to separate and fall down again. They are tossed about in this manner, and slightly beat or patted with the hands, for a considerable space of time. At length, when they become soft and flaccid, they are thrown in heaps, and *allowed to lie in this state for about an hour, or perhaps a little longer.* When examined at the end of this time, they appear to have undergone a slight change in colour, are soft and moist, and emit a fragrant smell.

The rolling process now commences. Several men take their stations at the rolling-table, and divide the leaves amongst them. Each takes as many as he can press with his hands, and makes them up in the form of a ball. This is rolled upon the rattan-worked table, and greatly compressed, the object being to get rid of a portion of the sap and moisture, and at the same time to twist the leaves. These balls of leaves are frequently shaken out, and passed from hand to hand until they reach the head workman, who examines them carefully to see if they have taken the requisite twist. When he is satisfied of this, the leaves are removed from the rolling-table and shaken out upon flat trays, until the remaining portions have undergone the same process. In no case are they allowed to lie long in this state, and sometimes they are taken at once to the roasting-pan.

The next part of the process is exactly the same as in the manipulation of green tea. The leaves are thrown into an iron pan, where they are roasted for about five minutes, and then rolled upon the rattan-table.

After being rolled, the leaves are shaken out, thinly, on sieves, and exposed to the air out of doors. A framework for this purpose, made of bamboo, is generally seen in front of all the cottages among the tea hills. The leaves are allowed to remain in this condition for about three hours: during this time the workmen are employed in going over the sieves in rotation, turning the leaves and separating them from each other. A fine dry day, when the sun is not too bright, seems to be preferred for this part of the operation.

The leaves having now lost a large portion of their moisture, and having beabout many chemical changes within the substance of the leaves themselves. The result of these changes is to produce the varied flavours, odours, and tastes by which different varieties of tea are more or less distinguished.

Second, That the treatment or mode of handling by which the leaves are converted respectively into green and black teas, is the cause of the different colours of these two main varieties. Thus, for

Green Teas.

1°. The leaves are roasted almost immediately after they are gathered.

2°. They are dried off quickly after the rolling process. The whole operation is speedy and simple.

Black Teas.

1°. They are allowed to be spread out in the air for some time after they are gathered.

2°. They are then further tossed about till they become soft and flaccid.

3°. They are now roasted for a few minutes, and rolled; after which, they are exposed to the air for a few hours in a soft and moist state.

4°. Lastly, they are dried slowly over charcoal fires.

It is by lengthened exposure to the air, therefore, in the process of drying, accompanied, perhaps, by a slight heating

come considerably reduced in size, are removed into the factory. They are put a second time into the roasting-pan for three or four minutes, and taken out and rolled as before.

The charcoal fires are now got ready. A tubular basket, narrow at the middle and wide at both ends, is placed over the fire. A sieve is dropped into the tube, and covered with leaves, which are shaken on it to about an inch in thickness. After five or six minutes, during which time they are carefully watched, they are removed from the fire and rolled a third time. As the balls of leaves comes from the hands of the rollers, they are placed in a heap until the whole have been rolled. They are again shaken on the sieves as before, and set over the fire for a little while longer. Sometimes the last operation—namely, heating and rolling—is repeated a fourth time; the leaves have now assumed a dark colour.

When the whole have been gone over in this manner, they are placed thickly in the baskets, which are again set over the charcoal fire. The workman now makes a hole with his hand through the centre of the leaves, to allow vent to any smoke or vapour which may rise from the charcoal, as well as to let up the heat, which has been greatly reduced by covering up the fires. The tea now remains over the slow charcoal fire, covered with a flat basket, until it is perfectly dry,—carefully watched, however, by the manufacturer, who every now and then stirs it up with his hands, so that the whole may be equally heated. The black colour is now fairly brought out, but afterwards improves in appearance. The after processes, such as sifting, picking, and refining, are carried on at the convenience of the workmen.

and fermentation, that the dark colour and distinguishing flavour are given to the black teas of commerce. The oxygen of the atmosphere acts rapidly upon the juices of the leaf during this exposure, and changes chemically the peculiar substances they contain, so as to impart to the entire leaf the dark hue it finally acquires. The precise nature, however, of these changes has not as yet been chemically investigated.

This action of the air does not appear sensibly to affect the weight of the tea obtained, as three pounds of the fresh leaves produce on an average about one pound of marketable tea of either kind. The teas intended for home consumption are not so highly dried as those which are prepared for exportation—(Dr. Bowring)—a circumstance which must affect the quality of the beverage they yield.

The produce of different districts varies in quality and flavour with the climate, the soil, and the variety of plant cultivated, as well as with the period at which the leaves are gathered, and with the mode of drying them. The finest tea of China grows between the 27th and 31st parallels of north latitude, on a low range of hills, which is an offshoot of the great chain of Pe-ling. The principal varieties of *black* tea are known by the names of Bohea, Congou, Campoi, Souchong, Caper, and Pekoe. Of these the bohea grows in the province of Fu-kian (Fokien). Pekoe, or pak-ho, means "white down" in Chinese, and consists of the first downy sprouts or leaf-buds of three-year-old plants. A very costly tea of this kind, known as the "Tea of the wells of the Dragon," is used only by persons of the highest rank in China, and is never brought to Europe. Caper is in hard grains, made up of the dust of the other varieties cemented together by means of gum. The *green* teas are known as Twankay, Hyson-skin, Hyson, Imperial, and Gunpowder. The hyson is grown in the province of Song-ho. The true imperial, known also, because of its excellence, as the flot-theac, seldom comes to Europe,—that which is usually sold under this name being really Chusan tea flavoured with the cowslip-coloured blossoms of the sweet-scented olive (*Olea fragrans*). The practice of scenting teas is very common, and various odoriferous plants are employed for the purpose in different parts of China.* It is remarked, however, by the dealers in tea, that the plantations which naturally yield a produce of a particularly-esteemed flavour are as limited in extent as the vineyards in Europe which are celebrated for particular kinds of wine. The price of tea varies, of course, with the variations in natural quality, being for some samples double or treble what is asked for others. But the average price at Canton is about 8½d. a-pound, so that the grower must sell it at 5d. or 6d. (MEYEN.)

Tea-leaves prepared as above-described have been in use as a beverage in China from very remote periods. Tradition speaks of it as early as the third century. The legend relates, "that a pious hermit, who, in his watchings and prayers, had often been overtaken by sleep, so that his eyelids closed, in holy wrath against the weakness of the flesh, cut them off and threw them on the ground. But a god caused a tea-shrub to spring out of them, the leaves of which exhibit the form of an eyelid bordered with lashes, and possess the gift of hindering sleep. A similar story is related concerning the introduction of coffee into Arabia. Both legends were probably invented long after the qualities of tea and coffee were known.

It was after the year 600 that the use of tea became

* Among these are mentioned the Olea fragrans, Chloranthus inconspicuus, Gardenia florida, Aglaia odorata, Mogorium sambac, Vitex spicata, Camellia sasanqua, Camellia odorifera, Illicium anisatum, Magnolia yulan, Rosa indica odoratissima, Murraya exotica, turmeric, oil of Bixa orellana, and the root of the Florentine Iris. With such a list before us, we cannot wonder that teas should exhibit great diversity in fragrance and flavour.

general in China, and early in the ninth century (810) it was introduced into Japan. To Europe it was not brought till about the beginning of the seventeenth century. Hot infusions of leaves had been already long familiar as drinks in European countries. Dried sage-leaves were much in use in England,* and are even said to have been carried as an article of trade to China by the Dutch, to be there exchanged for the Chinese leaf, which has since almost entirely superseded them. A Russian embassy to China also brought back to Moscow some carefully-packed green tea, which was received with great acceptance. And in the same century (1664) the English East India Company considered it as a rare gift to present the Queen of England with two pounds of tea!" †

The growth and consumption of tea are now really enormous. Mr. Ingham Travers estimates the total produce of the dried leaf in China alone at a million of tons, or 2240 millions of pounds! ‡ To this is to be added the tea of Japan, Corea, Assam, and Java. The produce of this latter island already goes far to supply the markets of Holland; and the introduction of the tea-plant into the hill-country of India promises to add largely to its future growth. The quantity of tea yielded by an acre of land is not stated in any book to which I have access; but if we take it at 600 lbs., which is probably a full estimate, the extent of land devoted to this branch of rural industry in China alone must be nearly 3½ millions of acres!

The consumption of tea in the United Kingdom in 1352 amounted to 55 millions of pounds (24,000 tons)—about

* Sage was in frequent use till after the middle of last century. In the life of Whitfield, it is stated, that, when in his fasting humours at Oxford, "he ate nothing but sage tea without sugar, and coarse bread." This was about 1730.

† *The Plant*, by SCHLEIDEN. Second Edition, p. 142.

‡ *A Few Words on the Tea Duties*. London, 1853

one forty-fifth part of the estimated produce of China. This is at the rate of 1 lb. 9 oz. per head of the population, and the consumption is rapidly on the increase. Among European nations tea is pre-eminently a British, Dutch, and Russian drink. Among the other nations of Europe, coffee and cocoa are more usual beverages than tea. This is strikingly illustrated by the fact, that while in 1835 about 36 millions of pounds of tea were consumed in the United Kingdom, only 200,000 lbs. were consumed in the kingdom of Prussia! The population of Prussia was then upwards of thirteen millions.

The effects of tea, as it is used in China, are thus described by Chinese writers: "Tea is of a cooling nature, and, if drunk too freely, will produce exhaustion and lassitude. Country people, before drinking it, add ginger and salt to counteract this cooling property. It is an exceedingly useful plant. Drink it, and the animal spirits will be lively and clear. The chief rulers and nobility esteem it; the lower people, the poor, and beggarly will not be destitute of it. All use it daily, and like it." Another writer says, "Drinking it tends to clear away all impurities, drives off drowsiness, removes or prevents headache, and it is universally in high esteem."*

The mode of using it in China is to put the tea into a cup, to pour hot water upon it, and then to drink the infusion off the leaves, and without admixture. While wandering over the tea districts of China, Mr. Fortune only once met with sugar and a tea-spoon.

The mode of making and drinking the infusion of tea probably does not alter its general effects upon the system. In China cold water is disliked, and considered as unwholesome, and therefore tea is taken to quench the thirst, which

* FORTUNE'S *Tea Districts of China*, vol. ii. p. 231.

it probably does best when drunk unmixed. The universal use, on the other hand, of sugar and cream or milk among us, probably arose from its being introduced here as a beverage among grown-up people whose tastes were already formed, and who required something to make the bitter infusion palatable. The practice thus begun has ever since continued, and, physiologically considered, is on the whole, I believe, an improvement upon the Eastern fashion.

The effects of tea as obtained and thus used among us are too familiarly known to require any detailed explanation. It exhilarates without sensibly intoxicating. It excites the brain to increased activity and produces wakefulness; thence its usefulness to hard students, to those who have vigils to keep, and to persons who labour much with the head. It soothes, on the contrary, and stills the vascular system, and hence its use in inflammatory diseases and as a cure for headache. Green tea, when taken strong, acts very powerfully upon some constitutions, producing nervous tremblings and other distressing symptoms, acting as a narcotic, and in inferior animals even producing paralysis.* Its exciting effect upon the nerves makes it useful in counteracting the effects of opium and of fermented liquors, and the stupor sometimes induced by fever.

In manufactured tea there are at least three active chemical substances, by the conjoined influence of which these effects are produced.

1°. *The volatile oil.*—When commercial tea is distilled with water there passes over a small quantity of a volatile oil, which possesses the aroma and flavour of the tea in a high degree. A hundred pounds of tea yield about one pound of this oil, and to this minute quantity of its volatile ingredient the value of tea in general estimation is in a great

* New tea in China is said to exhibit this narcotic quality in a high degree, and hence the Chinese rarely use tea before it is a year old.

measure due. Its special action upon the system has not yet, we believe, been scientifically investigated. But that it does exercise a powerful, and most likely a narcotic influence, is rendered probable by many known facts. Among these I mention the headaches and giddinesses to which tea-tasters are subject; the attacks of paralysis to which, after years, those who are employed in packing and unpacking chests of tea are found to be liable; and the circumstance already alluded to, that in China tea is rarely used till it is a year old, because of the peculiar intoxicating property which new tea possesses. The effect of this keeping upon tea must be chiefly to allow a portion of the volatile ingredients of the leaf to escape. And lastly, that there is a powerful virtue in this oil is rendered probable by the fact, that the similar oil of coffee has been found by experiment to possess narcotic properties.

This volatile ingredient does not exist in the natural leaf, but is produced during the process of drying and roasting already described.

2°. *The Theine.*—When dry finely-powdered tea-leaves are put upon a watch-glass, covered over with a conical cap of paper, and then placed upon a hot plate, a white vapour gradually rises from the leaves, and condenses on the inner side of the paper in the form of minute colourless crystals. If, instead of the leaves, a dried watery extract of the leaves be employed, the crystals will be obtained in greater abundance. These crystals consist of the substance known to chemists by the name of Theine or Caffeine. The teas of commerce contain, on an average, about two per cent. of this theine.—(STENHOUSE.) In some it is a little more. Certain green teas, according to Peligot, contain as much as six pounds in every hundred pounds of the dried tea; but so large a proportion as this is very rare.

Theine has no smell, and only a slightly bitter taste.

It has little to do, therefore, either with the taste or flavour of the tea from which it is extracted. It is remarkable, however, in three respects—

First, in containing a very large per centage of nitrogen, an element I have already spoken of as forming a large proportion of our common atmospheric air, and as distinguishing the gluten of wheat from the starch with which it is associated in the grain.* The composition of the dried theine is represented by the following numbers—

Carbon,	49.80
Hydrogen,	5.08
Nitrogen,	28.83
Oxygen,	16.29
	100

It contains, therefore, nearly three-tenths of its weight of nitrogen; a proportion which exists in only a very small number of other known substances.

Second.—Theine is remarkable as being present not only in Chinese tea, but also in Maté or Paraguay tea, in coffee, and in guarana—a substance prepared and used in Brazil in the same way as coffee. It is a very curious fact that, in countries so remote from each other, plants so very unlike as all these are should have been, by a kind of instinct as it were, selected for the same purpose of yielding a slightly exciting, exhilarating, and refreshing beverage; and that these plants, when now examined by chemists, should all be found to contain the same remarkable compound body which we call theine or caffeine. The selection must have been made by the independent discovery, in each country and by each people, that these several plants were capable of gratifying a natural constitutional craving, or of supplying a want equally felt by all.

* See The Air we breathe *and* The Bread we eat.

Third.—The observed effects of this substance, when introduced into the system, justify this conclusion, and form the third point which is worthy of remark in regard to it. It is known that the animal body, while living, undergoes constant decay and renovation. The labours of life waste it —the food introduced into the stomach renews it. That which is wasted passes off through the lungs and the kidneys, or is in other ways rejected from the body of the animal. The solid matters contained in the urine are in some degree a measure of this waste; and especially the quantity of urea and phosphoric acid it contains at different periods, is supposed to measure the comparative waste of the tissues at these different times. Now, the introduction into the stomach of even a minute proportion of theine—three or four grains a-day—has the remarkable effect of sensibly diminishing the absolute quantity of these substances voided in a day by a healthy man, living on the same kind of food, and engaged in the same occupation, under the same circumstances. This fact indicates that the waste of the body is lessened by the introduction of theine into the stomach—that is, by the use of tea. And if the waste be lessened, the necessity for food to repair it will be lessened in an equal proportion. In other words, by the consumption of a certain quantity of tea, the health and strength of the body will be maintained in an equal degree upon a smaller supply of ordinary food. Tea, therefore, saves food—stands to a certain extent in the place of food—while at the same time it soothes the body and enlivens the mind.

In the old and infirm it serves also another purpose. In the life of most persons a period arrives when the stomach no longer digests enough of the ordinary elements of food, to make up for the natural daily waste of the bodily substance. The size and weight of the body, therefore, begin to diminish more or less perceptibly. At this period tea comes in as a medicine to arrest the waste, to keep the body from falling away so fast, and thus to enable the less energetic powers of digestion still to supply as much as is needed to repair the wear and tear of the solid tissues.

No wonder, therefore, that tea should be a favourite, on the one hand, with the poor, whose supplies of substantial food are scanty—and on the other, with the aged and infirm, especially of the feebler sex, whose powers of digestion and whose bodily substance have together begun to fail. Nor is it surprising that the aged female, who has barely enough of weekly income to buy what are called the common necessaries of life, should yet spend a portion of her small gains in purchasing her ounce of tea. She can live quite as well on less common food, when she takes her tea along with it; while she feels lighter at the same time, more cheerful, and fitter for her work, because of the indulgence.

The quantity of three or for grains of theine, mentioned above, is contained in less than half an ounce of good tea,* and may be taken in a day by most full-grown persons, without unpleasant effects. But if twice this quantity, or eight grains a-day, be taken, the pulse becomes more frequent, the heart beats stronger, trembling comes on, and a perpetual desire to void urine. At the same time the imagination is excited, and, after a while, the thoughts wander, visions begin to be seen, and a peculiar state of intoxication comes on; all these symptoms are followed by, and pass off in, a deep sleep. The effects of strong tea, therefore,—and especially of old teas, and such as are peculiarly rich in theine—are to be ascribed in great part to the overdose of this substance which has been introduced into the stomach.

3°. *The Tannin* or tannic acid.—If tea be infused in hot water in the usual manner, and the infusion be poured

* An ounce of good tea contains about ten grains of theine.

into a solution of common green copperas (sulphate of iron), the mixture will become black. Or if it be poured into a solution of glue or isinglass (gelatine), it will render the solution turbid or muddy, and cause a greyish precipitate to fall. These appearances show that the tea contains an astringent substance, known to chemists by the name of tannin or tannic acid. This substance is so called, because it is the ingredient which, in oak bark, is so generally employed for the tanning of leather.

To this tannic acid tea owes its astringent taste, its constipating effect upon the bowels, and its property of giving an *inky* infusion with water which contains iron. It forms from 13 to 18 per cent. of the whole weight of the dried tea-leaf, and is the more completely extracted the longer the tea is infused. The tannic acids, of which many varieties are known to chemists, though naturally colourless, have all a tendency to become dark-coloured when exposed to the air. This is one reason why the same leaves, when dried quickly, will give a *green*, and when dried more slowly, a *black* tea, as has been described by Mr. Fortune.

What is the full and precise action of this tannin uponthe system as we drink it in our tea, or whether it contributes in any degree to the exhilarating, satisfying, or narcotic action of tea, is not yet known. That it does aid even in the exhilarating effect which tea produces, is rendered very probable by the fact, that a species of tannin is the principal ingredient in the Indian betel-nut, which is so much chewed and prized in the East, and which is said to produce a kind of mild and agreeable intoxication.*

4°. *The Gluten.*—The three substances already described may be considered as the really active constituents of the tea-leaf as it is usually employed. But it is an interesting

* See THE NARCOTICS WE INDULGE IN.

fact, that the leaf contains a large proportion of that nutritive ingredient of plants to which the name of gluten * is given. This substance forms as much as one-fourth of the weight of the dry leaves; so that if we chose to eat them in mass, they would prove as nutritious as beans or peas. This is seen by the following table, which exhibits the composition of beans and of tea-leaves as they are severally brought to market:—

	Tea-leaves.	Beans.
Water,	5	14
Starch, gum, &c.,	27†	48
Gluten,	20 to 25	24
Fat,	8	2
Tannic acid,	15†	—
Husk or woody fibre,	20†	10
Ash,	5	2
	100	100

Of this large per-centage of gluten, the water in which we usually infuse our tea, extracts very little; and hence we throw away in the waste leaves a large proportion of the common nutrition they contain. It has been recommended, therefore, as an improved method of infusing tea, that a pinch of soda should be put into the water along with it. The effect of this would be, that a portion at least of the gluten would be dissolved, and the beverage in consequence made more nutritious. The method of preparing the brick tea adopted among the Mongols and other Tartar tribes, is believed to extract the greater part of the nutriment from the leaf. They rub the tea to fine powder, boil it with the alkaline steppe-water, to which salt and fat have been added, and pour off the decoction from the sediment. Of this liquid they drink from 20 to 40 cups a-day, mixing it first with

* See THE BREAD WE EAT.

† The ingredients marked with a † are very variable in quantity in the tea-leaf.

milk, butter, and a little roasted meal. But even without meal, and mixed only with a little milk, they can subsist upon it for weeks in succession.

The effect of the tea in this way of using it seems to be twofold. *First*, it directly nourishes by the gluten and milk or meal it contains; and, *second*, it makes this food go farther, through the waste-retarding influence of the theine, which the boiling thoroughly extracts.

But the most perfect way of using tea is that described, I think, by Captain Basil Hall, as practised on the coast of South America, where tea-leaves, after being exhausted by infusion, are handed round the company upon a silver salver, and partaken of by each guest in succession. The exhilarating effects of the hot liquid are in this practice followed by the nutritive effects of the solid leaf. It is possible that this practice may refer to the Paraguay tea, so extensively used in South America; but in either case the merit of it is the same.

The four substances above mentioned are the most important ingredients of the tea-leaf. It contains besides, as is shown by the table given above, a large proportion of starch and gum, some of which will, of course, be extracted by boiling water, and will give a certain nutritive value to the infusion. Tea, however, varies in composition with the mode of drying, with the age of the plant and of the leaf, with the season in which it is gathered, and even with the variety of shrub on which it has grown. Hence the proportion of the whole leaf which is extracted by boiling water varies much both in kind and quantity. The genuine green teas, which are usually prepared from the young leaves, yield more of the lighter coloured—the black teas more of the darker coloured, ingredients. And even of teas of the same colour and name in the market, different samples yield to boiling water very different proportions of soluble matter. Two samples of souchong, for example, examined by Davy and Lehmann, respectively gave, to boiling water, from a hundred parts—

$82\frac{1}{3}$ parts to Davy,
$15\frac{1}{3}$ „ to Lehmann.

It is obvious, therefore, that the value of tea as a beverage, in so far as this depends on the proportion of soluble matter it contains, differs very much. We usually judge of the quality of a tea by its aroma, and by the flavour and colour of the infusion it yields; and these, in the main, are good guides: but chemistry indicates that, as in the case of opium, some weight ought also to be attached to the proportion of soluble ingredients it contains and readily yields to boiling water.

It is necessary to mention, before concluding my remarks upon tea, that, in addition to the substances which it naturally contains, others are sometimes added by way of adulteration to the teas of commerce. This is especially the case with the green teas, which are not all prepared by simply drying quickly the natural leaf as already described, but are often artificially coloured by the addition of blue, white and yellow colouring substances. Mr. Fortune, who saw the colouring performed in China, thus describes the process:—

"The superintendent having taken a portion of Prussian blue, threw it into a porcelain bowl not unlike a mortar, and crushed it into a very fine powder. At the same time a quantity of gypsum was burned in the charcoal fire which was then roasting the tea. This gypsum having been taken out of the fire after a short time, readily crumbled down, and was reduced to powder in the mortar. The two substances thus prepared were then mixed together, in the proportion of four of gypsum to three of Prussian blue, and formed a light blue powder, which was then ready for use.

"This colouring matter was applied to the teas during

the last process of roasting.* About five minutes before the tea was removed from the pans, the superintendent took a small porcelain spoon, and with it he scattered a portion of the colouring matter over the leaves in each pan. The workmen then turned the leaves rapidly round with both hands, in order that the colour might be equally diffused. To 14 lbs. of tea about 1 oz. of colouring matter was applied.

"During this part of the operation the hands of the workmen were quite blue. I could not help thinking that if any green-tea drinkers had been present during the operation, their taste would have been corrected and improved.

"One day an English gentleman in Shanghae, being in conversation with some Chinese from the green-tea country, asked them what reasons they had for dyeing the tea, and whether it would not be better without undergoing this process. They acknowledged that tea was much better when prepared without having any such ingredients mixed with it, and that they never drank dyed teas themselves; but remarked that, as foreigners seemed to prefer having a mixture of Prussian blue and gypsum with their tea, to make it look uniform and pretty, and as these ingredients were cheap enough, the Chinese had no objections to supply them, especially as such teas always fetched a higher price!" *

Mr. Fortune describes the blue substance employed as Prussian blue; and Mr. Warrington's experiments † show that, until the last few years, this substance was very generally in use in China for giving an artificial colour to teas. More recently, however, it appears that indigo has been substituted, in consequence, probably, of the injurious effects which European writers have described the Prussian blue as likely to produce on the constitution of green-tea drinkers.

* FORTUNE'S *Tea Countries of China*, vol. ii. p. 69.

† See *Transactions of the Chemical Society.*

The quantity of either substance employed, however, is so minute that, without justifying the adulteration, I think it unlikely that any serious consequences can have followed from it. The indigo is probably harmless; but supposing it to be Prussian blue, the quantity added to the green tea is about one grain to the ounce; and this is already diluted to a pale tint with white clay, so as not to contain more than a third, or probably a fourth, of a grain of pure Prussian blue. This quantity in an ounce of tea is, I think, but little to be dreaded; nevertheless the practice ought to be discouraged and abandoned.*

Less doubt exists as to the pernicious qualities of an adulterated tea largely manufactured by the Chinese under the name of Lie tea. This consists of the sweepings and dust of the tea-warehouses cemented together with rice-water and rolled into grains. It is made either black to imitate caper, or green to resemble gunpowder, and is manufactured professedly for the purpose of adulterating the better kinds of tea.

Genuine tea yields only 5 or 6 per cent. of ash when burned, being the proportion of mineral matter naturally contained in the leaf. The lie teas leave from 37 to 45 per cent. of ash, consisting chiefly of sand and other impurities. These adulterated teas are imported into this country to the extent of half a million pounds' weight every year! In this, as in similar cases, the poorest classes, who can least afford it, are the greatest sufferers from the fraudulent

* It is easy to determine whether indigo or Prussian blue is the colouring matter of these adulterated teas. If a portion of the tea be shaken with cold water and thrown upon a bit of thin muslin, the fine colouring matter will pass through the muslin, and settle to the bottom of the water. When the water is poured off, the blue matter may be treated with chlorine or a solution of chloride of lime. If it is bleached, the colouring matter is indigo. If potash makes it brown, and afterwards a few drops of sulphuric acid make it blue again, it is Prussian blue.

introduction of the lieing mixture into the teas they buy. Among the low dealers the lie tea is known by the name of *dust and gum.*

2°. Maté, or Paraguay tea, though not used over so large an area as the Chinese tea, is as much the passion of the Brazilians and their neighbours, in Southern America, as the latter is of the nations of north-eastern Asia. It is prepared from the dried leaves of Brazilian holly (*Ilex Paraguayensis*),—(fig. 30)—is said to have been in use among the Indians from time immemorial,—has been drunk by all classes in Paraguay since the beginning of the seventeenth century, and is now consumed by "almost the whole population of South America." The leaf of this tree is 4 or 5 inches long, and after being dried it is rubbed to powder before it is infused. The dried leaf has much of the aroma of some varieties of Chinese tea, and the infusion has a pleasant odour, and an agreeable bitter taste. In the state in which it is commonly used in South America, it is more exciting than China tea, producing a kind of intoxication, and by excessive use leading even to *delirium tremens.*

Fig. 30.

Ilex Paraguayensis--Paraguay holly (Paraguay Tea-plant.)
Scale, 1 inch to 10 feet.
Scale for leaf, 1 inch to 4 inches.

The tree which yields the Yerba (or plant *par excellence*), as this tea is called, does not appear to be an object of culture. It grows spontaneously, in extensive natural plantations, amid the forests of Paraguay. The principal Yerbals, or woods of this tree, are situated in the neighbourhood of a small town called Villa-Real, about fifteen hundred miles above Assumption, on the Paraguay river. They are scattered about, however, in various other localities upon the rich tract of country which extends between the rivers Paraná and Uruguay. Permission to gather the leaves is granted by the government to certain merchants, in return for a considerable money payment. These merchants fit out parties of men, chiefly Indians, for the purpose of collecting the Yerba, and at the proper season proceed to the forests. When in the course of their journey they come to a Yerbal, or growth of maté trees, sufficiently extensive to make it worth while to halt and collect the leaves, they begin by constructing a long line of wigwams, which they cover with the broad leaves of the banana and palm. Under these they expect to pass nearly six months. An open space is then prepared, of which the soil is beaten with heavy mallets until it becomes hard and smooth. Over this is erected a kind of arch, made of hurdles, called a *Barbagua*, upon which the Yerba branches are placed. Beneath these a large fire is kept up till the foliage is thoroughly dried and roasted, without being scorched or suffered to ignite. The hard floor is then swept clean, the dried branches are laid upon it, and the now brittle leaves beaten off with sticks, which partly reduce them to powder. They are then crammed and beaten into sacks made of damp hides, which, when sewed up and left to dry, become in a few days as hard as stone. In these sacks, weighing about 200 lb., the maté is well preserved. The labour of collecting the Yerba, in the midst of these tropical forests, is very severe, and it is said to have been very fatal to Indian life. Many of the Creoles and Mestizos even assert that the Paraguayans have exterminated the poor Indians by compelling them to the labour of collecting this plant.

From the smallest shrubs the finest tea is obtained; but from the same kind of leaves different qualities are procured, according to the mode of preparation, and the kind of weather which prevails. Three principal kinds, however, are prepared and sold in South America under the names of caa-cuys, caa-miri, and caa-guaza—the prefix caa signify ing the leaf itself. The *first* is prepared from the half-expanded buds: it will not keep, and its consumption is entirely confined to Paraguay. The *second*, from the leaf carefully picked and stripped from the nerves before roasting, as was done by the Jesuits. And the *third*, from the entire foliage, roasted as above described, without any preparation. The two latter varieties are not only used largely in the country of Paraguay, but are exported as far as Lima and Quito.—(HOOKER.)

We have no data from which to calculate the total amount of maté consumed either in the whole of South America or in Paraguay alone. But it must be very large, since the quantity exported from the latter country is about 50,000 quintals, or 5,600,000 lb. a-year. It loses in virtue and flavour, however, and its aromatic bitterness diminishes by exportation and keeping, so that the infusion is drunk in perfection only on the spot where the leaves are gathered and newly dried.

In Brazil, a variety of maté called Gongonha is in use. It is prepared from the leaves of two other species of holly, the *Ilex gongonha* and the *Ilex theezans;* but I do not know to what extent. In Chili also, a tea called Paraguay tea, but different from the maté, is prepared from the leaves of the *Psoralea glandulosa*, and in Central America another variety from those of the *Capraria biflora*.

The use of the maté is very frequent, as well as very universal in South America. At every meal, and at every hour of the day, it is drunk. It has acquired the name of Maté from that of the vessel or cup in which it is infused, and from which it is drunk. Hot water is poured upon the powdered leaf, then a lump of burned sugar, and sometimes a few drops of lemon juice are added. The infusion is sucked through a tube, *bombilla*, often made of silver, which is open at one end, and has a perforated bulb or strainer at the other (fig. 31). The cup is passed from hand to hand, the same cup, and often the same tube, serving a whole party. The leaves will bear to be steeped or watered three times, and the infusion is drunk off quickly, as it soon becomes black if allowed to stand.

Fig. 31.

Maté or cup, and Bombilla or tube.

"Persons who are fond of maté consume about an ounce a-day. In the mining districts it is most universally taken, experience having shown that fermented liquors are there prejudicial to health.* The Creoles in South America are passionately fond of the beverage, and never travel without a supply of the leaf, which they infuse before every meal, and sometimes much oftener, never tasting food unless they have first drunk their maté." †

* A maxim of the Jesuits was, "*En pais caliente, aguardiente; en pais frio, agua frio*"—in the warm country, brandy; in the cold country, water.

† HOOKER's *London Journal of Botany*, vol. i. p. 89.

Numerous virtues are ascribed to this favourite beverage. It possesses many of the good qualities of our Chinese tea, while, like opium, it is said to calm the restless, and to arouse the torpid. As is the case with opium also, the habit of using it becomes a kind of second nature, so that to give it up, or even to diminish the customary quantity, is almost impossible. On the other hand, long indulgence, or an immoderate consumption of it, is apt to induce diseases similar to those which follow the excessive use of ardent spirits. It differs both from Chinese tea and from opium in acting upon the kidneys and moving the bowels.

The chemistry of the maté leaf is but imperfectly understood. From being rarely met with in Europe, it has not been much examined by chemists, yet we are sufficiently acquainted with the nature of its constituents to be able to account for its most striking effects. Thus—

First,—Like Chinese tea, it contains a volatile oil, which is produced during the roasting of the leaf, gives it a peculiar, agreeable aroma, gradually escapes from it by keeping, and upon which a portion of its narcotic virtue depends. This is shown by the facts already stated, that the tea becomes less valuable when long kept, or carried to great distances, and that it is only drunk in perfection near the Yerbal, where it is collected and prepared.

Second,—Dr. Stenhouse has shown that this leaf also contains theine, the vegetable principle which we have described as existing in Chinese tea, and as producing remarkable effects upon the system when introduced into the stomach. The proportion, however, is somewhat less than in Chinese tea, amounting in the varieties hitherto examined in Europe, to about 1¼ per cent.

Third,—Paraguay tea contains a large proportion of a peculiar, astringent acid, analogous to the tannin or tannic acid. For this reason, the fresh leaves are used in Brazil by the dyers. It is probably the presence of this substance in the infusion which causes it to blacken so rapidly when exposed to the air, and makes it necessary to drink it off as soon as it is made. Were it poured out into cups, as is done with Chinese tea, the liquid would become black and repulsive before the eyes of the drinker. Hence the reason for the peculiar mode of sucking it through a tube, which is practised in South America, and which at first seems so peculiar to Europeans. And,—

Lastly,—Like the Chinese leaf, it contains also nutritious gluten. Of this substance about 10 per cent. is present in the dried maté, of which only a small proportion dissolves when the tea is infused. The benefit of this ingredient, therefore, is experienced only where the infused leaf is subsequently eaten, as is the case, it is said, in some parts of South America.

An exact analysis of Paraguay tea has not yet been made, so that we are still in the dark as to its precise composition; but it is both interesting and remarkable to find, so far, a great similarity between the Chinese and the South American leaf. Both contain the same active ingredients, and both, though belonging to very different tribes of plants, have been selected to serve the same remarkable physiological purposes. How came tribes so remote, and so little civilized, to stumble upon this happy selection?

3°. COFFEE-TEA.—Attention has lately been drawn to the use of the leaf of the coffee-tree as a substitute for that of the tea-tree. In 1845 Professor Blume of Leyden, who had travelled much in Java, made known in Holland that this leaf was so used in the Eastern Archipelago, and recommended it for use in Europe. Subsequently it was made known in this country by Professor Brand; * and at the Great Exhibition in 1851, Dr. Gardner showed speci-

* *Chemistry*, p. 108.

mens of prepared coffee-leaves, announced at the same time that they contained *theine*, and suggested that they should be substituted for our ordinary tea.

These, along with other circumstances, have drawn the attention of Eastern merchants to the subject, and it appears from various communications which have recently been made public, that the use of coffee-leaves in this way is an old practice in the Eastern Archipelago. In the Dutch island of Sumatra especially, prepared coffee-leaves form "the only beverage of the whole population, and, from their nutritive qualities, have become an important necessary of life."

The leaves are roasted over a clear, smokeless, bamboo fire, till they become of a brownish-buff colour. They are then separated from the twigs, the bark of which, after a second roasting, is rubbed off and used along with the leaves. In this state they have an extremely fragrant odour, resembling that of a mixture of tea and coffee. When immersed in boiling water, they give a clear brown infusion, which, with sugar and cream, forms an agreeable beverage. Mr. Ward, who has been many years settled at Pedang in Sumatra, thus narrates his experience in regard to the use of the coffee-leaf in that island:—

"The natives have a prejudice against the use of water as a beverage, asserting that it does not quench thirst, or afford the strength and support the coffee-leaf does. With a little boiled rice and infusion of the coffee-leaf, a man will support the labours of the field in rice-planting for days and weeks successively, up to the knees in mud, under a burning sun or drenching rain, which he could not do by the use of simple water, or by the aid of spirituous or fermented liquors. I have had the opportunity of observing for twenty years the comparative use of the coffee-leaf in one class of natives, and of spirituous liquors in another—the native Sumatrans using the former, and the natives of British India settled here, the latter; and I find that, while the former expose themselves with impunity to every degree of heat, cold, and wet, the latter can endure neither wet nor cold for even a short period, without danger to their health.

"Engaged myself in agriculture, and being in consequence much exposed to the weather, I was induced several years ago, from an occasional use of the coffee-leaf, to adopt it as a daily beverage, and my constant practice has been to take two cups of a strong infusion, with milk, in the evening, as a restorative after the business of the day. I find from it immediate relief from hunger and fatigue. The bodily strength is increased, and the mind left for the evening clear and in full possession of its faculties. On its first use, and when the leaf has not been sufficiently roasted, it is said to produce *vigilance;* but I am inclined to think that, where this is the case, it is rather by adding strength and activity to the mental faculties, than by inducing nervous excitement. I do not recollect this effect on myself except once, and that was when the leaf was insufficiently roasted.

"As a beverage the natives universally prefer the leaf to the berry, giving as a reason that it contains more of the bitter principle, and is more nutritious. In the lowlands, coffee is not planted for the berry, not being sufficiently productive; but, for the leaf, the people plant it round their houses for their own use. It is an undoubted fact that everywhere they prefer the leaf to the berry."*

He adds further, that while the culture of the coffee plant, for its fruit, is limited to particular soils and more elevated climates, *it may be grown for the leaf wherever, within the tropics, the soil is sufficiently fertile.* This is a very important fact, and, should the leaf come into general use, will no doubt lead to the introduction of new forms of

* *Pharmaceutical Journal,* vol. xiii, p. 208.

husbandry in many tropical regions, from which the coffee-tree, as a profitable article of culture, has been hitherto excluded. The Brazilian government is said to be directing its attention to the subject, and shipments of prepared coffee-leaves are announced to have been already made from that country to Europe. At present the price of prepared leaves in Sumatra is about 1½d. a pound; and they may be packed of good quality, for the European market, for 2d. a-pound.

In regard to the constituents of the dried coffee-leaf, the agreeable aroma emitted shows that, like Chinese tea, it contains a volatile oil, which will probably act upon the system like the similar oils of tea and coffee. It has been proved also to contain theine to the extent of 1¼ per cent.—(Stenhouse)—and an astringent acid closely resembling that which is found in Paraguay tea. Both of these are present in it in larger proportion than in the coffee-bean; and hence, probably, the reason why the leaf is preferred to the bean by the natives of Sumatra. These, with about 13 per cent. of gluten and some gum, are all the ingredients yet found in the leaf. But the presence of these substances proves it to be so similar to the tea-leaf in composition, as to lead to the belief that it may be successfully substituted in common use for the Chinese tea. And this conclusion is supported by the wakefulness which is said to be produced by the infusion of coffee-leaves, and by the bodily refreshment it is found to yield, by the directly nutritive power which the leaves possess, and by the general favour they have found in the estimation of the people of Sumatra.

To boiling water the dried coffee-leaves yield about 39 per cent. of their weight—as much as is taken up by water from the most soluble varieties of the coffee-bean, and more than is yielded by Chinese tea. In this property, therefore, the leaf of the coffee-tree is also equal to the bean.

4°. Labrador tea is the name given in North America to the dried leaves of the *Ledum palustre* and the *Ledum latifolium* (fig. 32). These plants grow on the borders of the swamps, and along the heathy shores of the mountain lakes in the colder regions of that continent. The leaves are gathered and used in the stead of Chinese tea—the narrower leaved plant (*L. palustre*), according to Dr. Richardson, giving tea of the better quality. Both varieties are very astringent, and possess a narcotic, soothing, and exhilarating quality. This latter is so strong that in the north of Europe (Sweden and Germany) these plants are secretly employed by fraudulent brewers to give headiness to beer. They have not been examined chemically; but from the above facts we may infer that, besides a variety of tannin, to which they owe their astringency, they contain an active narcotic principle, more powerful, probably, than the theine of the tea-leaf, to which their peculiar, exhilarating, and stupefying effects are due. It is possible also, that, in the cold northern climates of Sweden and Labrador, the effects of such a narcotic substance may be less sensibly felt than under our milder skies.

Fig. 32.

Ledum palustre—The Marsh Ledum, or Labrador Tea.
The undermost flower and leaf represent those of
Ledum latifolium—The Labrador Tea, or broad-leaved Ledum.
Scale, 1 inch to 2 feet.
Leaves and flowers nearly natural size.

5°. Abyssinian tea, called in its native country Khat or Chaat, is very extensively cultivated in Shoa and the adjoining regions, and is in general use among the inhabitants, just as tea is in China. It consists of the dried leaves of the *Catha edulis*, a species of small tree which is allied to the *Sageretia theezans*, from which the poorer classes of Chinese prepare an inferior kind of tea. In a light gravelly soil the plant attains a height of 12 feet. The leaves are plucked in the dry season, and well dried in the sun. In Abyssinia they sell at 1d. or 2d. a-pound.* They are either chewed, boiled in milk, or infused in boiling water, and, by the addition of honey, yield a pleasant beverage. They have much resemblance to Chinese tea, both in their qualities and their effects. They are bitter to the taste, possess exhilarating properties, and dispel sleep if used to excess.

The leaves of this plant are also used green. Forskäll states that the Arabs eat them green because of their property of preventing sleep. To such a degree do they exhibit this influence, that a man who chews them may stand sentry all night without feeling drowsiness. They are also regarded as an antidote to the plague; and the Arabs believe that the plague cannot appear in places where the tree is cultivated. Botta adds to these qualities that, when fresh, the leaves are very intoxicating.†

This North African tea appears to be very extensively cultivated and used, though less so now than in ancient times; but we have no means of estimating the absolute quantity which is grown and consumed. We are entirely ignorant, also, I believe, of its exact chemical history, and do not yet know whether it belongs to the class of plants in which theine exists. Its relation to the *Sageretia theezans* of China renders this not unlikely.

Many other plants, of which the chemistry is unknown, are used in various countries as more or less perfect substitutes for Chinese tea. Thus, the name

Tasmanian tea is given to the dried leaves of various spe-

* HARRIS—*Highlands of Ethiopia*, vol. ii. p. 423.
† LINDLEY—*Vegetable Kingdom*, p. 587.

cies of Melaleuca and Leptospermum, belonging to the order of the Myrtaceæ, which are collected in Australia, and used by the colonists instead of Chinese tea. These trees are commonly called tea-trees, and the large tracts of country which are covered with them, *tea-tree flats.* The leaves of various species of Correa also, which belong to the Rutaceæ, and especially of the *Correa alba*, are collected and used for the same purpose. The leaves of *Acæna sanguisorba*, a plant allied to the Rosaceæ, and which abounds everywhere in Tasmania, are said to be an excellent substitute for tea. In the same eastern region the leaves of the *Glaphyria nitida*, another of the Myrtaceæ—called by the Malays the Tree of Long Life, affords at Bencoolen, in Sumatra, a substitute for tea.

Faham tea, again, is the name given in Mauritius to the dried leaves of the *Angræcum fragrans*—a fragrant orchid. The infusion of these leaves is exceedingly pleasant to the smell, and is drunk to promote digestion, and in certain diseases of the lungs. Its fragrance is owing to the presence of *coumarin*, the odoriferous principle of the Tonka bean and of mellilot, described in a subsequent chapter.* The leaf does not contain theine, and it is not therefore to be classed in its virtues and uses with the Chinese and Paraguay teas.

Besides all these we have North American substitutes for the China leaf, distinguished by the names of Appalachian tea, Oswego tea, Mountain tea, and New Jersey tea. We have a Mexican tea, a Brazilian tea,—the aromatic *Capitaô da matto*,—a Santa Fé tea, an Indian, Toolsie tea and many others. Of the chemistry of all these substitutes we know next to nothing. I have therefore embodied in the following table nearly all the information we possess regarding them:—

* See THE ODOURS WE ENJOY.

LIST OF SUBSTITUTES FOR CHINESE TEA AND MATÉ.

Name of the Plant.	Natural order.	Where collected and used.	Name given to it.
Hydrangea thunbergii, . .	Hydrangeaceæ.	Japan.	Ama tsja or Tea of Heaven.
Sageretia theezans,	Rhamnaceæ.	China.	?
Ocymum album,	Labiatæ.	India.	Toolsie tea.
Catha edulis, . .	Celastraceæ.	Abyssinia.	Khat or Chaat.
Glaphyria nitida,	Myrtaceæ.	Bencoolen (flowers used).	Tea-plant and Tree of Long Life.
Correa alba, . .	Rutaceæ.	New Holland.	Tea plants, and Tasmanian tea.
Acæna sanguisorba, . .	Sanguisorbiaceæ.	Do.	
Leptospermum scoparium, and L. thea, . .	Myrtaceæ.	Do.	
Melaleuca scoparia, and M. genistifolia, . .	Myrtaceæ.	Do.	
Myrtus ugni,	Myrtaceæ.	Chili.	Substitutes for Paraguay tea.
Psoralea glandulosa,	Leguminosæ.	Do.	
Alstonia theaformis,	Styracaceæ.	New Granada.	Santa Fé tea.
Capraria biflora.	Scrophulariaceæ.	Central America.	?
Lantana pseudothea,	Verbenaceæ.	Brazil.	Capitaô da matto.
Chenopodium ambrosioides, .	Chenopodiaceæ.	Mexico and Columbia.	Mexican tea.
Viburnum cassinoides, . .	Caprifoliaceæ.	North America.	Appalachian tea.
Prinos glaber, .	Aquifoliaceæ.	Do.	
Ceanothus Americanus, . .	Rhamnaceæ.	Do.	New Jersey tea (medicinal).
Gaultheria procumbens, .	Ericaceæ.	Do.	Mountain tea.
Ledum palustre, Ledum latifolium,	Ericaceæ.	Do.	Labrador tea, or James' tea.
Monarda didyma, M. purpurea, .	Labiatæ.	Do.	Oswego tea.
Angræcum fragrans, . .	Orchidiaceæ.	Mauritius.	Bourbon or Faham tea.
Micromeria thea-sinensis, .	Labiatæ.	France.	?
Stachytarpheta jamaicensis,	Verbenaceæ.	Austria.	Brazilian tea.
Prunus spinosa, ⅓ mixed with ⅔ Fragaria collina, or F. vesca, .	Drupaceæ. Rosaceæ.	Northern Europe.	Sloe and Strawberry tea, one of our best substitutes for Chinese tea.
Salvia officinalis,	Labiatæ.	Do.	Sage tea.

I pass over numerous other plants which in Europe have been tried as substitutes for tea, without, however, coming into any general use, except here and there as adulterations. It is possible that some of those above mentioned may hereafter be discovered to contain the theine and other valuable constituents of the true tea-leaf, and may be both cultivated and advantageously used in its stead. As an adulteration, the leaves of *Epilobium angustifolium* are sometimes mixed with tea to the amount of 25 per cent.

CHAPTER VIII.

THE BEVERAGES WE INFUSE.

THE COFFEES.

Coffee used in Abyssinia from time immemorial—Its introduction into Europe—Consumption in the United Kingdom, in Europe, and in the whole world—Varieties of coffee, and prices in the London market—Effects of the infusion of coffee—It exalts the nervous life, and lessens the waste of the system—Constituents of coffee—The volatile oil, its production, mercantile value, and effects on the system—The tannic acid, the theine or caffeine, and the gluten—Composition of tea and coffee compared—Loss of weight in roasting coffee—Proportion of the roasted bean taken up by water very variable—Substitutes for coffee—Seeds of the water-iris, of the Turkish kenguel, of the roasted acorn, of roasted corn and pulse, of roasted roots, and especially of chicory—The chicory plant and root—How the root is prepared for use—Gives a fictitious appearance of strength to coffee—Active ingredients in chicory—The empyreumatic oil, and the bitter principle—Its effects on the system—Mode of detecting chicory in coffee—Adulterations of chicory.

II. The Coffees.—The name of coffee is given to a beverage prepared from the seeds of plants roasted, ground, and infused in boiling water. The seeds of the Arabian coffee-tree are most largely used for this purpose, but various other seeds are more or less extensively employed in a similar way.

1°. Arabian coffee.—The tree which produces this seed is said to be indigenous to the countries of Enárea and Cáffa in southern Abyssinia. In these districts the coffee-tree grows like a wild weed over the rocky surface of the country. The roasted seed or bean has also been in use as a beverage in Abyssinia generally, from time immemorial, and is at the present day extensively cultivated in that country. In Persia it is known to have been in use as early as the year 875. From Abyssinia it was introduced into Arabia in the beginning of the fifteenth century, when it partly superseded the older chaat, or Abyssinian tea. About the middle of the sixteenth century it began to be used in Constantinople, and in spite of the violent opposition of the priests, became an article of general consumption. In the middle of the seventeenth century (1652), the first coffee-house was opened in London by a Greek named Pasqua; and twenty years after, the first was established in Marseilles. Since that time both the culture and consumption of coffee have continually extended. It has become the staple produce of important colonies, and the daily and most cherished drink of probably more than a hundred millions of men!

The consumption in the United Kingdom in 1852 amounted to 35 millions of pounds, of which upwards of 20 millions were brought from Ceylon, 4 millions from Jamaica, and 8 millions from Costa Rica and Brazil. On the Continent it is much more generally used than among ourselves. The total European consumption was estimated a few years ago at 75 thousand tons, or 168 millions of pounds, valued at 4½ millions sterling. It probably approaches now to 200 millions of pounds. The entire weight of coffee raised over the whole world is guessed at about 600 millions of pounds.

The quality of raw coffee does not appear to depend so much on the mode of collecting and drying it as that of tea does. Soil and climate are the circumstances which chiefly affect its commercial value. The flavour and quality of the beverage prepared from it depend very much, however, upon the manner of roasting the bean, and of subsequently preparing the infusion.

In the London market the coffees of different countries are arranged, as to quality and price, in the following order.

The third column of this table shows the quantity of each sort consumed in the United Kingdom in 1852 :—

	Wholesale price per cwt.	Consumed in 1852.
Ceylon, native, ...	46s. to 47s.	20,500,000 lb.
Do. Plantation, . .	52s. to 80s.	
East India, ...	48s. to 78s.	1,600,000 —
Costa Rica and Brazil, . .	50s. to 70s.	6,700,000 —
Jamaica, ...	50s. to 100s.	4,000,000 —
Mocha (ungarbled), ...	50s. to 60s.	1,800,000 —
Do.	68s. to 90s.	
Other sorts, ...		400,000 —
		35,000,000 lb.

The Arabian or Mocha coffee is small, and of a dark yellow colour. The Javan and East Indian are larger, and of a paler yellow. The Ceylon, West Indian, and Brazilian have a bluish or greenish-grey tint.

Fig. 33.

Coffea Arabica—Arabian-Coffee-Tree. Scale, 1 inch to ten feet. Scale for leaf, 1 inch to 2 inches.

The coffee-tree (fig. 33) when in good health, and full grown, attains a height in some countries not exceeding 8 or 10, but in others averaging from 15 to 20 feet, and is covered with a dark, smooth, and shining foliage. It is sown in nurseries —transplanted when about six months old—in three years comes into full bearing, and in favourable circumstances will continue to bear for twenty years. It delights in a dry soil and a warm situation. On dry and elevated spots the berries are smaller, and have a better flavour; but berries of all sizes improve in flavour or *ripen* by keeping, The small berries of Arabia will ripen in three years, but the worst coffee produced in America will, in from ten to fourteen years, become "as good, and acquire as high a flavour, as the best we now have from Turkey."—(Ellis.)

The sensible properties and effects of coffee, like those of tea, are too well known to require to be stated in detail It exhilarates, arouses, and keeps awake; it allays hunger to a certain extent, gives to the weary increased strength and vigour, and imparts a feeling of comfort and repose. Its physiological effects upon the system, so far as they have been investigated, appear to be, that, while it makes the brain more active, it soothes the body generally, makes the change and waste of matter slower, and the demand for food in consequence less.* All these effects it owes to the conjoined action of three ingredients, very similar to those contained in tea. These are a volatile oil produced during the roasting—a variety of tannic acid, which is also altered during the roasting—and the substance called theine or caffeine, which is common to both tea and coffee.

First, The Volatile Oil.—When the coffee-bean is gathered and dried in the air it has little smell, and only a slight-

* The influence of coffee in retarding the waste of the tissues—as indicated by the quantity of phosphoric acid, common salt, and urea discharged under its influence in a day—was shown by estimating the proportions of each of these ingredients voided in his urine by the same person, in the same circumstances, when he drank coffee and when he took none.

	Urine, containing	Phosphoric acid.	Common salt.	Urea.
		grammes	grammes	grammes
H. S., without coffee, voided	1635 c. c.	4.421	9.865	31.298
With coffee from 1¼ oz. of beans	2005 "	3.001	8.819	21.888
Difference, . .	+ 370 c. c.	— 1.420	— 1.046	— 9.410

In this experiment, while the absolute quantity of urine discharged in the twenty-four hours was increased more than one-fifth, the absolute quantities of urea and of phosphoric acid contained in the urine were diminished one-third. That is to say, the change or waste of matter, as indicated by the contents of the urine, was diminished to that extent by the influence of the coffee. And the natural inference from this is, that the occupation of the individual being the same, the necessary demand for ordinary food would be lessened in a somewhat corresponding degree.

ly bitter and astringent taste. As with the tea-leaf, it is during the roasting of coffee that the much prized aroma and the greater part of the taste and flavour are brought out or produced. In tea, as we have seen, the proportion of volatil oil amounts to about one pound in a hundred of the dried leaf, but in roasted coffee it rarely amounts to more than one in fifty thousand! And yet on the different proportions of this oil which they severally contain, the aroma and the consequent estimation in the market of the different varieties of coffee in a great measure depend. A higher aroma would make the inferior Ceylon, Jamaica, and East Indian coffees nearly equal in value to the finest Mocha; and if the oil could be bought for the purpose of imparting this flavour, it would be worth in the market as much as £100 sterling an ounce!—(PAYEN). How it comes—by what slow chemical change within the bean, that is, that coffee of the most inferior quality so ripens by keeping as at length to yield, on roasting, a coffee equal to the finest Mocha, we do not as yet know. The oil is formed during the roasting by the action of the heat on some substance present in the natural bean, probably in small quantity only. It is possible that by prolonged keeping this substance is itself formed in the inferior qualities of coffee; so that when roasted after the keeping a larger quantity of the valuable aromatic oil is formed in the bean.

The effect of this volatile oil of coffee upon the system has been made the subject of direct experiment. When roasted coffee is distilled with water this oil passes over, and by drinking the distilled water and oil together its effects may be ascertained. Julius Lehmann found in this way that is has an effect in retarding the waste of the tissues quite equal to that of caffeine itself.* It produces also an agree-

* The relative effects of the volatile oil of coffee, of caffeine, and of the infusion of coffee, made in the ordinary way, upon the same individual (G. M.) in his usual

able excitement, and a gentle perspiration, dispels the sensation of hunger, and moves the bowels. In its exhilarating action upon the brain it affects the imagination less than the reasoning powers.

These effects followed when the quantity of oil yielded by two ounces of coffee was taken in a day. If this dose was doubled, violent perspiration came on, with sleeplessness and symptoms of congestion.

It appears, therefore, that the volatile empyreumatic oily constituents of roasted coffee, though present only in minute quantity, exercise a powerful influence upon the animal economy, exciting to greater activity both the vascular and nervous systems, and yet retarding the waste of the tissues in as great a degree as the caffeine itself, which the infusion of coffee usually contains. This activity of the oil of coffee justifies us in concluding, as I have already said, that the similar oil produced in tea by the roasting takes a similar share in the effects which the infusion of tea as a beverage produces.

Second, The Astringent Acid.—The raw coffee contains about 5 per cent. of an astringent acid—the caffeine or caffee-tonic—which does not blacken a solution of iron, as the

state of health, and when consuming the same food in kind and quantity, were found by Julius Lehmann to be as follows:—

	Urine. { containing {	Phosphoric acid.	Common salt.	Urea.
		grammes	grammes	grammes
Without coffee, he voided daily	1444 c. c.	4.140	9.363	27.232
With 4 grains caffeine, do.,	1928 "	3.768	9.546	24.088
With empyreumatic oil from 2 oz. of beans, . . .	1789 "	3.479	10.307	20.271
With coffee, from 1½ oz. of beans,	1512 "	3.105	6.951	20.695

In all trials the quantity of the urine was increased; but, in all, the total quantity of saline matter contained in the urine was lessened. The urea, as shown in the last column, was diminished most by the empyreumatic oil, but the waste of phosphoric acid and common salt more by the coffee itself, which contained both, than by either of the ingredients when used alone.

infusion of tea does, but renders it green,* and does not precipitate solutions of gelatine. This acid is changed to some extent during the roasting, but still retains a portion of its astringent properties, and contributes in some degree to the effects which the infusion of coffee produces upon the system.

It will be observed that the proportion of this astringent principle contained in coffee is much less than is contained in tea. Hence it is not sufficient to retard the action of the bowels as tea does, especially when associated with the empyreumatic volatile oil, which, as we have seen, has a positive tendency to move them. To the same result the large per-centage of fat contained in coffee may also contribute.

Third, The Theine, or Caffeine as it is also called, exists in different proportions in different varieties of coffee. It varies in the coffee usually employed in this country from three quarters of a pound to one pound in the hundred—(STENHOUSE),—though according to some experimenters, three or four pounds in the hundred occur in certain varieties of coffee. By rubbing common roasted coffee in a mortar with a fifth of its weight of slaked lime, and then boiling the mixture in alcohol, about half a per-cent. of theine may be readily extracted. Weight for weight, therefore, tea yields about twice as much theine as roasted coffee does to the water in which it is infused. But as we generally use a greater weight of coffee than we do of tea in preparing our beverages, a cup of coffee of ordinary strength will probably contain as much theine as a cup of ordinary tea.

The influence which this ingredient of the several beverages has in producing the effects we experience from the

* Many varieties of the astringent, so-called tannic acids are found in plants—that which exists in tea has much resemblance to the tannin of the oak, while those of coffee, of Paraguay tea, and the heaths (Ericaceæ), form another class of acids—having much resemblance to one another, but differing in their properties from the tannic acid of the oak.

use of them, has already been explained when treating of the effects of tea.

But the coffee-bean contains also about thirteen per cent. of nutritious gluten, which, as in the case of tea, is very sparingly dissolved by boiling water, and is usually thrown away in the insoluble dregs of the coffee. Among some of the Eastern nations, the custom prevails of drinking the *grounds* along with the infusion of the coffee: in these cases the full benefit is obtained from all the positively nutritive matter which the roasted coffee contains.

The composition of unroasted coffee, compared with the average composition of the tea-leaf as it comes to Europe, is nearly as follows:—

	Tea. (Mulder.)	Coffee. (Payen.)
Water, ...	5	12
Gum and sugar . .	21	15½*
Gluten, . . .	25	13
Theine, ...	½	¾
Fat and volatile oil, . .	4	13
Tannic acid, . . .	15	5
Woody fibre, . . .	24	34
Ash, ...	5½	6¾
	100	100

The proportion of theine in both tea and coffee, it will be recollected, is somewhat variable.

Coffee swells by roasting, but loses in weight, and assumes a brown colour more or less dark. These changes vary, however, with the degree of roasting. Thus—

Roasted to a	It loses in weight	And gains in bulk
Reddish brown, .	15 per cent.	30 per cent.
Chestnut brown, .	20 per cent.	50 per cent.
Dark brown, . .	25 per cent.	50 per cent.

* According to Dr. Stenhouse, coffee contains as much as eight per cent. of cane sugar.

The aroma is most agreeable when the heat is not greater than is sufficient to impart a light brown colour to the bean. When the roasting is carried too far a disagreeable smell gradually mingles with the esteemed aroma, and lessens the value of the product.

The quantity of the coffee-bean which is taken up by water is nearly the same before and after roasting. It is nearly the same also in some samples, whether they be much or little roasted. It differs, however, very much in different samples. Thus three experimenters found that water extracted from the samples of roasted coffee they examined, the following proportions per cent. :—

	Payen.	Cadet.	Lehmann.
Reddish brown,	37.0	12⅓	21⅛
Chestnut brown,	37.1	18½	—
Dark brown,	37.2	23⅓	—

Some infusions of coffee, therefore, even when roasted to the same extent, contain three times as much of the solid substance of the coffee as others do. But we have no experiments upon the comparative effects which infusions so differing have upon the constitution of the drinkers. It is observed that some natural waters give a stronger and better flavoured coffee than others; and this has been traced, as in Prague, to the presence of alkaline matter in those which give the most agreeable infusion. Hence, to obtain a more uniformly strong and well-flavoured coffee, it is recommended to add a little soda to the water with which the infusion is made. About forty grains of dry, or twice as much of crystallized carbonate of soda, are sufficient for a pound of coffee.

The chemical changes caused by the roasting, are the production of the active empyreumatic oil, and of a brown, bitter substance, the chemical properties of which, and its action upon the system, still remain to be investigated. They are produced from the soluble part of the raw bean, but by what chemical changes is not yet known. In conclusion, it is proper to state that coffee is reputed to possess important medicinal virtues. The great use of coffee in France is supposed to have abated the prevalence of the gravel. In the French colonies, where coffee is more used than in the English, as well as in Turkey, where it is the principal beverage, not only the gravel, but the gout, is scarcely known. Among others, also, a case is mentioned of a gentleman who was attacked with gout at twenty-five years of age, and had it severely till he was upwards of fifty, with chalk stones in the joints of his hands and feet; but the use of coffee then recommended to him completely removed the complaint.*

It has not been determined to which of the constituents of coffee this curative action is due, or whether it is the same in all constitutions. These points are worthy of careful experimental investigation.

2°. OTHER COFFEES.—Besides the real *Coffea Arabica*, other species of the coffee-plant are grown in various countries, and yield a useful marketable bean. Thus, in Silhet and Nepaul, the *Coffea Benghalensis* is cultivated; on the coast of Mosambique, the *Coffea Mosambicana;* on the coast of Zanguebar, the *C. Zanguebaria;* and in the Mauritius, the *C. Mauritiana.* The seed of the last of these tastes disagreeably sharp and bitter, and sometimes causes vomiting, yet it is in some places cultivated instead of the *Coffea Arabica.* It is possible that these so-called different species may, like the varieties of the tea-plant, be only differently modified forms of the same original species.

But, besides the fruit of the different coffee-plants, numerous other vegetables have, in different countries, been proposed or used as substitutes for Arabian coffee. A successful substitute must contain, like coffee, a fragrant aro-

* *Pharmaceutical Journal*, vol. xiii. p. 330.

matic principle, a bitter principle, and an astringent principle. These properties are found more or less satisfactorily—

a. In the roasted seeds of *Iris pseudacoris* (yellow water-iris), which are said to approach very near to coffee in quality.

b. In the seeds of a Goumelia, called in Turkey Kenguel, which were shown at the Great Exhibition as extensively cultivated in the Kair-ar-eh and Komah, where they are roasted, ground, and used as coffee.

c. In the roasted acorn, which is said to be much used on the Continent under the name of acorn coffee.

d. In the cicer or chick-pea roasted; in beans, rye, and other grains; in nuts, almonds, and even in wheaten bread, when roasted carefully.

e. In the seeds of Broom (*Spartium scoparium*), and in the dried and roasted berries of the *Triosteum perfoliatum* (Caprifoliaceæ). In the West Indies, the seeds of several species of Psychotria (Cinchonaceæ); in Soudan, those of Dura and Nitta (*Inga biglobosa*); among the African negroes, those of *Parkia* (*Africana*); and among the Tonguses, those of a species of Hyoscyamus—are all employed as substitutes for coffee.

f. In the dried and roasted roots also of many plants. The carrot and turnip are used for this purpose, but more commonly the roots of the common goose-grass (*Galium aparine*), especially in Ireland; while those of the dandelion (*Leontodon taraxacum*) and of chicory are extensively employed both in this country and on the Continent. In none of these roots, however, has the characteristic principle, theine, been discovered, and none of them, therefore, can serve physiological purposes as the seeds of our common coffee.

Yet one of these roots (chicory) has already, in other

countries, crept into extensive use, and among ourselves is at present rapidly rising in public estimation. At first it was only mixed with pure coffee as an adulteration by fraudulent dealers. But this practice extended itself so widely, that, for the defence both of the honest dealer and of the public, the sale has been legalised, and much chicory in the unmixed state is now bought and used instead of or along with genuine coffee. As one of the recognised beverages we *now* infuse, therefore, the plant deserves a brief notice in this place.

Fig. 34.

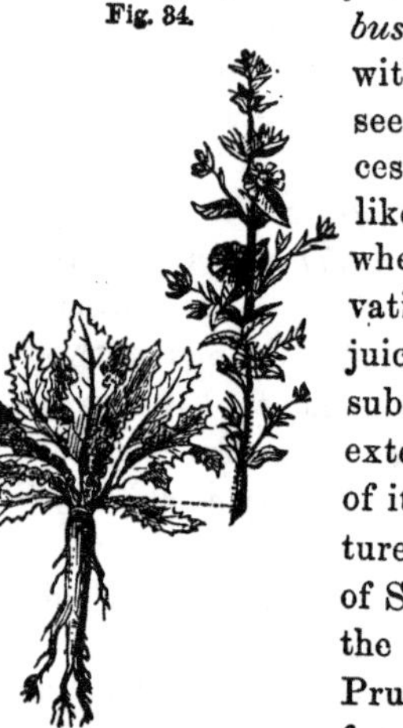

Cichorium intybus—The Chicory plant.
Scale, half-inch to a foot.

3°. Succori, chicory or wild endive (*Cichorium intybus*), fig. 34, is a native weed, which, with its large pale-blue flowers, is seen scattered about in numerous places. It has a large white parsnip-like tap-root, which increases in size when the plant is subjected to cultivation. This root abounds in a bitter juice, which has led to its use as a substitute for coffee. The plant is now extensively cultivated for the sake of its root. In this country the culture is chiefly confined to the counties of Surrey, Bedford, and York. On the Continent it is largely grown in Prussia, Belgium, and France. The foreign is considered greatly superior to that of English growth, and is largely imported into this country, chiefly through Hamburg and Antwerp.

The root is taken up before the plant shoots into flower, is washed, sliced, and dried; it is then roasted till it is of a chocolate colour. Two pounds of lard are roasted with each hundredweight, and the root loses in roasting from 25 to

30 per cent. When ground and exposed to the air, it becomes moist and clammy, increases in weight, and acquires a distinct smell of liquorice, and a sensibly sweet *first* taste. It possesses in no degree the pleasant aroma which recommends the genuine roasted coffee. When infused, even in cold water, it imparts to it a dark colour, and a sweetish-bitter taste. To many the addition of a little of this bitter liquid to the infusion of genuine coffee appears an improvement—a remarkable illustration of the creation of a corrupt taste by an adulteration, which taste demands afterwards the continuance of the adulteration to satisfy its own craving. The bitter substance however, is not considered unwholesome. Very many bitter substances of this kind possess a tonic property, and it is not unlikely that the bitter of chicory may be among the number.

But the use of chicory appears to have originated from other causes than the discovery, or even the supposed presence, of a tonic property in its bitter ingredient. A little of the roasted chicory gives as dark a colour to water, and as bitter a taste, as a great deal of coffee, and hence it was originally introduced into the coffeehouses for a purpose akin to that which takes *Cocculus indicus* into the premises of the fraudulent brewer. It gave colour and taste to the beverage of the drinker, and at the same time saved the expensive coffee of the seller. The public taste gradually accommodated itself to the fraudulent mixture; it became by-and-by even grateful to the accustomed palate; and finally a kind of favourite necessity to the lovers of *bitter coffee*. How far circumstances are gradually giving to the infusion of chicory, in some countries, the character of a national beverage, may be judged of from the facts, that in 1845 the quantity of chicory imported into this country was estimated at 2000 tons, or 4½ millions of pounds, and it has since largely increased; that the quantity of the dried root consumed in

France amount already to 12 millions of pounds a-year; and that in some parts of Germany the women are becoming regular chicory-topers,* and are making of it an important part of their ordinary sustenance.

The active ingredients in roasted chicory are, *first*, the empyreumatic volatile oil; this is produced during the roasting, and though not so fragrant, this oil probably exercises upon the system some of the gently-exciting, nerve-soothing, and hunger-staying influence of the similar ingredients contained in tea and coffee; and, *second*, the bitter principle. When taken unmixed, this substance is to many, while they are unaccustomed to it, not only disagreeable, but nauseous in a high degree. It may, however, like many other bitter principles, possess, as I have said, a tonic or strengthening property. Taken in moderate quantities, these ingredients of chicory are probably not injurious to health; but by prolonged and frequent use they produce heartburn, cramp in the stomach, loss of appetite, acidity in the mouth, constipation, with intermittent diarrhœa, weakness of the limbs, tremblings, sleeplessness, a drunken cloudiness of the senses, &c. &c. At the best, therefore, chicory is a substitute for coffee to which only those to whom the price is an object ought to have recourse.

The simplest way of detecting an admixture of chicory in coffee, is to put the powder in cold water. Chicory gives a coloured infusion in the cold while coffee does not, and by the depth of the colour the proportion of chicory may be guessed at. The presence of coffee in chicory is ascertained by boiling the supposed mixture with quicklime, filtering, evaporating to dryness, adding sulphuric acid and peroxide of manganese, and gently heating, when a substance called *kinon* will sublime, if coffee is present.

* "Cichorien-Kaffee-Schwelgerinnen."—STRUMPF, *Die Fortschritte der Angewandten Chemie.*

The infusion or decoction of a suspected mixture may be tested also by salts of peroxide of iron. The infusion of chicory is brownish yellow, and becomes only a little darker when such a salt of iron is added, giving no precipitate. The infusion of coffee is of a brown colour, becomes green when the iron solution is added, and gives a brownish-green precipitate.

Another reason why the use of chicory should be avoided by those who can afford to buy pure coffee, is found in the fact, that pure chicory is as difficult to be met with in the market as unadulterated coffee. Venetian red is very commonly employed to impart to the chicory a true coffee colour; and it is curious to observe how the practice of adulteration extends itself from trade to trade. The coffee-dealer adulterates his coffee with chicory to increase his profits—the chicory-maker adulterates his chicory with Venetian red, to please the eye of the coffee-dealer; and, lastly, the Venetian-red manufacturer grinds up his colour with brick-dust, that by his greater cheapness, and the variety of shades he offers, he may secure the patronage of the trade in chicory!

CHAPTER IX.

THE BEVERAGES WE INFUSE.

THE COCOAS.

Cocoa, ancient use of, in Mexico.—Brought to Europe by the Spaniards.—The tree and its fruit.—Varieties in the market.—Quantity imported into this country.—Manufacture of the bean.—Cocoa nibs.—Cocoa of commerce.—Chocolate.—Constituents of cocoa.—The volatile oil.—The peculiar bitter principle, theobromine.—The large proportion of fat which characterises cocoa.—The starch and gluten.—Its general composition compared with that of milk.—It forms a most nutritious beverage.—Substitutes for cocoa.—The earth-nut and the guarana of Brazil.—Decoction of cocoa nibs not so nutritious.—The cocoa husk or "miserable;" importation of and beverage from.—General view of the chemistry of the infused beverages.—Summary of their physiological action.—Concluding reflections.—Prison dietaries.

III. The Cocoas, as I have said, are more properly soups or gruels than simple infusions. They are prepared from certain oily seeds, which are first ground to a pulp by passing them between hot rollers, and are then diffused through boiling water for immediate use.

1°. The Mexican cocoa is the seed of the *Theobroma cacao* (fig. 35). This is a small but beautiful tree, with bright dark-green leaves, which is a native of the West Indies and of the central regions of America. It grows spontaneously in Mexico, and on the coast of Caraccas and forms whole forests in Demerara.

Fig. 35.

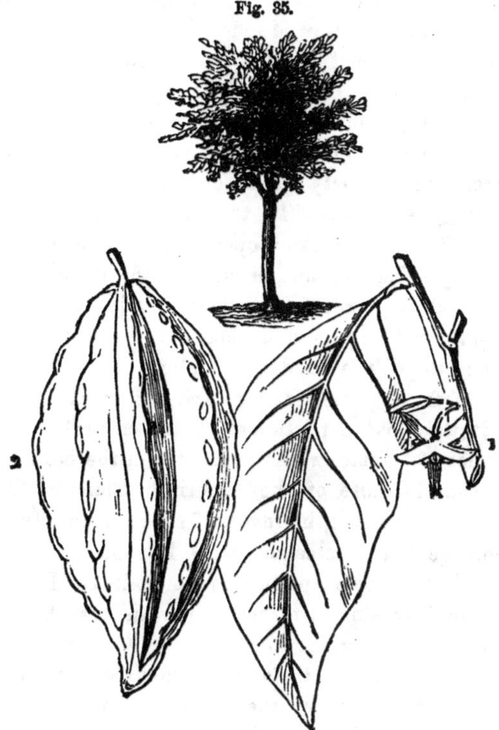

Theobroma cacao—The Cacao, "Cocoa," or Chocolate tree.
Scale, 1 inch to 10 feet.
1, Leaf and flower.—2, Fruit or pod.
Scale, 1 inch to 2 inches.

When the Spaniards first established themselves in Mexico, they found a beverage prepared from this seed in common use among the native inhabitants. It was known by the Mexican name of Chocollatl, and was said to have been in use from time immemorial. It was brought thence to Europe by the Spaniards in 1520, and has since been introduced more or less extensively as a beverage into every civilised country. Linnæus was so fond of it that he gave to the tree the generic name of Theobroma—Food of the Gods.

The fruit of the tree, which, like the fig, grows directly from the stem and principal branches, is of the form and size of a small oblong melon or thick cucumber (*see* fig. 36). It contains from six to thirty beans or seeds, imbedded in rows in a spongy substance, like that of the water-melon. When ripe, the fruit is plucked, opened, the seeds cleaned from the marrowy substance, and dried. In the West Indies they are immediately picked for market; but in the Caraccas they are put in heaps, and covered over, or sometimes buried in the earth till they undergo a slight fermentation, before they are finally dried and picked for market. By this treatment they lose a portion of their natural bitterness and acrimony of taste, which is greater in the beans of the mainland than in those of the American islands. The cocoa of Central America is, however, of superior quality, or at least is more generally esteemed in the European markets than that which is grown in the West Indies. It still retains a greater degree of bitterness, and this may be one reason for the preference given to it.

The cocoa of Trinidad is the variety chiefly consumed in this country. The quality of the mainland cocoas which come to the English market from Bahia and Guayaquil for example, has hitherto been always inferior. The reason of this has been, that, until the recent alteration of the tariff, the duty on British province cocoa was 1d. a-pound, and five per cent. additional; while on foreign cocoa it was 2d. a-pound, and five per cent. This difference was equal to one-fourth or one-fifth of the whole price of the cocoa; and, therefore, while it brought to our markets the best qualities produced in Trinidad and in our other colonies, it excluded all foreign cocoas but those which were of such inferior quality that, after paying this heavy duty, they could still be sold as low as the produce of our own plantations. The more choice varieties were sent to the markets of Mexico, Spain, France, and Italy, in which countries the beverages prepared from the cocoa-bean are more popular and in more general use than among ourselves. Indeed, they have never been favourites among us, nor has the consumption of cocoa kept pace even with the increase of our population. Thus the importation in—

1840 was	. . . .	3,499,746 lbs.
1842 „	. . . .	3,172,255 „
1852 „	. . . .	3,400,000 „

so that for twenty years the quantity imported yearly into the United Kingdom has been nearly stationary. By the recent alteration of the tariff, however, the duty on foreign cocoa has been reduced to a penny a-pound, the same as on British plantation cocoa. All qualities, therefore, will now come to us under equal advantages, and we may expect both that the article will be cheapened in the market, and that the consumption of it will largely increase.

The cocoa-bean of commerce is brittle, of a dark brown colour internally, eats like a rich nut, and has a slightly astringent but decidedly bitter taste. This bitterness is more decided in the South American or mainland varieties. In preparing it for use, it is gently roasted in an iron cylinder, in the same way as coffee is roasted, till the aroma appears to be fully developed, when it is allowed to cool. The bean is now more brittle, lighter brown in colour, and both the natural astringency and the bitterness are less perceptible than before. It is manufactured for the market in one or other of three principal ways. *First*, The whole bean after roasting is beat into a paste in a hot mortar, or is ground between hot rollers adjusted for the purpose. This paste, mixed with starch, sugar, and other similar ingredients in various proportions, forms the common cocoa, rock cocoa, soluble cocoa, &c., of the shops. These are often gritty from the admixture of earthy and other matters which adhere to the husk of the beans. *Second*, The bean is deprived of its husk, which forms about 11 per cent. of its weight, and is then crushed into fragments. These form the cocoa nibs of the shops, and are the purest state in which cocoa can usually be obtained from the retail dealer. *Third*, The bean, when shelled, is ground at once into a paste, by means of hot rollers, mixed with sugar, and seasoned with vanilla, and sometimes with cinnamon and cloves: this paste forms the long-known chocolate.

When prepared, it is also used in three different ways. *First*, The chocolate is made up into sweet cakes and bonbons, and is eaten in the solid state as a nutritious article of diet, containing in a small compass much strength-sustaining capability. *Second*, The chocolate or cocoa is scraped into powder, and mixed with boiling water or boiling milk, when it makes a beverage, somewhat thick, but agreeable to the palate, refreshing to the spirits, and highly nutritious. *Third*, The nibs are boiled in water, with which they form a dark-brown decoction, which, like coffee, is poured off the insoluble part of the bean. With sugar and milk this forms an agreeable drink, better adapted for persons of weak digestion than the consumption of the entire bean. Another variety of the cocoa beverages, and which may be called cocoa-tea, is prepared by boiling the husks of the bean in water, with which they form a brown decoction. This husk is usually ground up with the ordinary cocoas, but it is always separated in the manufacture of the purer chocolates. Hence in the chocolate manufactories it accumulates in large quantities, which are imported into this country from Trieste and other Italian ports, under the name of "miserable." Here the husk is partly ground up in the inferior cocoas, and is partly despatched to Ireland, where it is said to yield a wholesome and agreeable beverage to the poorer classes.

Besides the exhilarating and sustaining properties which it possesses in common with tea and coffee, cocoa, in its more common forms, is eminently nutritious. Its active or useful ingredients are the following:—

First, The volatile oil, to which its aroma is due, and which is produced during the roasting. The proportion of this oil which is contained in the roasted bean has not yet been determined, but it is no doubt very small. Its action on the system is probably similar to that of the odoriferous oils produced by the same process in tea and coffee.

Second, A peculiar principle, resembling the theine of tea and coffee, though not identical with it. Like theine, it is a white crystalline substance, which has a slightly bitter taste, and contains a large per-centage of nitrogen. It is called by chemists *theobromine*, from the generic name of the cocoa tree; and its composition, compared with that of theine, is as follows:—

	Theine.	Theobromine.
Carbon,	49.80	46.43
Hydrogen,	5.08	4.20
Nitrogen,	28.83	35.85
Oxygen,	16.29	13.52
	100	100

It is richer in nitrogen, therefore, even than theine; and as nearly all vegetable principles, rich in nitrogen, of which the influence upon the system has been examined, are found to be very active, the same is inferred in regard to theobromine. And further, its analogy in chemical properties to theine leads to the belief that it exercises a similar exhilarating and soothing, hunger-stilling and waste-retarding effect with the latter substance. The benefits experienced from the use of cocoa are due, in part at least, therefore, to the theobromine it contains. The proportion of this substance in the cocoa-bean is small, but it has not yet, I believe, been rigorously determined. It exists, also, in sensible quantity in the husk of the bean. The decoction obtained by boiling the husk in water, will not, therefore, be wholly devoid of useful ingredients, or of good effect.

Third, The predominating ingredient in cocoa, and the one by which it is most remarkably distinguished from tea and coffee, however, is the large proportion of fatty matter known as cocoa-butter which it contains. This amounts to upwards of one-half the weight of the shelled or husked bean. Consumed in either of its more usual forms, therefore, cocoa is a very rich article of food, and for this reason it not unfrequently disagrees with delicate stomachs. It is in some measure to lessen the sense of this richness, that sugar, starch, and fragrant seasonings are so generally ground up with the roasted bean in the manufacture of cocoa and chocolate.

Fourth, It contains also a large proportion both of starch and gluten,—substances which, as we have elsewhere seen, form the staple constituents of all our more valuable varieties of vegetable food. The average composition of the entire bean, when deprived of its husk, is nearly as fol lows :—

Water, ...	5
Starch, gum, &c., ...	22
Gluten, &c., ...	17
Oil (cocoa-butter), with a little theobromine, ...	56
	100

This composition reminds us of the richest and most nutritive forms of vegetable food; and especially of the oily seeds and nuts with which cattle are fed and fattened. Of all the varieties of human food, however, it has the closest resemblance to milk. Thus, dried milk (milk evaporated to dryness), and the dry cocoa-bean, consist respectively of—

	Milk.	Cocoa-bean.
Casein or gluten, ...	35	18
Fat, ...	24	55
Sugar or starch, &c., ...	37	23
Ash, or mineral matter, ...	4	4
	100	100

It is rich, therefore, in all the important nutritious principles which are found to co-exist in our most valued forms of ordinary food. It differs from milk chiefly by the greater proportion of fat which it contains, and hence it cannot be used so largely without admixture as the more familiar milk. When mixed with water, however, as it is usually drank, it is more properly compared with milk than with infusions of little direct nutritive value, like those of tea and coffee. And, on the other hand, it has the great advantage over milk, over beef-tea and other similar beverages, that it contains the substance theobromine, and the volatile empyreumatic oil. Thus it unites in itself the exhilarating properties of tea with the strengthening and ordinary body-supporting qualities of milk. The cocoa, as shown in the above table, is richer in fat, the milk in casein. Hence probably has arisen the practice of making milk-cocoa, in which the constituents of the one ingredient dovetail into and assuage the influence of those of the other. The large proportion of oil it contains justifies also, as fitting it better for most stomachs, the practice of mixing or grinding up the cocoa with sugar, flour, or starch, in the preparation of cocoa-paste or chocolate. Both practices are indeed skilful chemical adjustments, made without chemical knowledge, as the results of long and wide experience. And, lastly, the general composition of the beans shows that, in chocolate cakes and comfits, when faithfully prepared, there should reside, as experience has also shown to be the case, much nutritive virtue, and the means, both of supporting the bodily strength, and of sustaining the nervous energy reduced into comparatively small compass.

2°. BRAZILIAN COCOA, or Guarana.—In Brazil the seeds of the *Paullinia sorbilis* are collected, prepared, and used in the same way as those of the *Theobroma cacao.* They are usually described by travellers as a variety of coffee—but the seeds, like the cocoa-bean, are pounded and made into cakes, which are known as Guarana bread. When used, these cakes are mixed with water, as we do with the cakes of cocoa or chocolate, and the mixture is sweetened and drank. To what extent this article is prepared and consumed in Brazil, I have not been able to ascertain. It is a fact of great interest in regard to this substance, and one which shows it to have a true place among the beverages of which we are now treating, that like tea and coffee it has been found to contain theine, and is, therefore, capable of exercising upon the system an influence similar to that which is experienced by those who use these two favourite beverages.

3°. OTHER COCOAS.—The substances, as yet known, which can be employed in the place of, or as substitutes for, Mexican cocoa, are comparatively few in number. To fit them for this purpose, they must contain an odoriferous principle of some degree of fragrance, abundance of fat, and a considerable amount of ordinary nutriment. Oily seeds and nuts are almost the only vegetable productions from which beverages resembling cocoa have anywhere been manufactured. Among these the earth-nut (*Arachis hypogæa*), a kind of oily underground pea, is roasted in South Carolina, and then prepared and used in the same way as chocolate. In Spain, the root of the *Cyperus esculentus*, or earth-chestnut, is roasted and used as a substitute for both coffee and chocolate, but especially for the latter, which is much consumed in Spain. These are all the professed substitutes for the cocoa-bean with which I am acquainted. Neither of the two last-mentioned, however, contains a bitter principle rich in nitrogen, of the nature of the theobromine of the true cocoa, or of the theine contained in guarana. They can never, therefore, be employed effectively to replace the Mexican cocoa.

As adulterating materials, the substances chiefly employed by fraudulent manufacturers of cocoa and chocolate, are the husks of the bean, starch, sugar, fat, ground roots, and red ochre.

Before I leave this subject, it may interest the reader if I briefly sum up what appears to be the actual state of our knowledge regarding the chemistry and physiology of the beverages we infuse.

First, As to the chemistry of the various leaves and seeds we have mentioned, it appears that, when roasted and ready for use, they all contain,—

a. A volatile, odoriferous, aromatic oil, which does not exist in the fresh leaf or seed, but is produced or developed during the roasting. In tea this oil is most abundant, in coffee probably next, and in cocoa least in quantity. In the teas (Chinese and Paraguay), and in roasted coffee, the quantity and activity of this oil appear to diminish by keeping. In raw coffee, on the other hand, the power of developing this oil by roasting is greater the longer the bean is kept or allowed to ripen.

b. A peculiar, bitter, crystallisable principle, containing much nitrogen, and exerting a specific action on the system. In the teas, in coffee, and in guarana, this principle is theine, which contains 29 per cent. of nitrogen; in cocoa it is theobromine, which contains 36 per cent. of nitrogen. Weight for weight, the average qualities of tea contain about twice as much theine as the average qualities of coffee, but in both it varies between 1 and 5 per cent. as extremes. In cocoa the proportion of theobromine has not been determined. In well-roasted coffee, and in chicory, another bitter principle, which is soluble, uncrystallisable, and free from nitrogen, is produced during the roasting. The quantity and properties of this substance have not been determined.

c. A variety of tannin or tannic acid, which gives their astringency to the infusions prepared from all these substances. Of this ingredient the teas contain most, coffee next, and cocoa the least. The tannin of Chinese tea gives a black, that of maté and of coffee a green, with solutions containing iron.

d. A nutritious substance resembling the gluten of wheat or the fibrin of beef. In the tea-leaf this ingredient is most abundant, in cocoa next, while coffee contains the least. It dissolves but sparingly in water, and is therefore generally lost to the consumer when only the infusion is drank. The full benefit of this ingredient is obtained only when the tea-leaves are eaten, when the coffee grounds are taken along with the infusion, or when the whole material is made into a beverage, as in the usual modes of preparing cocoa and chocolate.

e. A quantity of fat, which in cocoa forms more than half the whole weight of the bean, in coffee one-eighth, and in tea only 3 or 4 per cent. The presence of so large a proportion of fat gives a peculiar character to cocoa, rendering it most nutritious, especially when made with milk, to those whose stomachs will bear it, but making it less suitable at the same time to persons of weak digestive powers.

Of the infusions themselves which are yielded by the different varieties of tea, maté and coffee, it is to be observed that they vary in strength with the sample employed. Of some teas and coffees, boiling water will extract and dissolve as much as one-third of the whole substance; of others, not more than one-sixth. The proportions of the several ingredients above-mentioned which the infusions we prepare are likely to contain, must therefore be very variable and uncertain.

Second, As to the physiology of these beverages, or their action on the system, it appears—

a. Generally, that they all exert a remarkable influence on the activity of the brain—exalting, so to speak, the nervous life; and yet they do so in a way different from opium or ardent spirits, since they act as antidotes to the narcotic influence of the one, and relieve the intoxication produced by the other.

b. They all soothe the vascular or corporeal system, allay hunger, retard the change of matter, and diminish the amount of bodily waste in a given time; and if this waste must, in the healthy body, be constantly restored in the form of ordinary food, this diminution of the waste is equivalent to a lessening of the amount of food which is necessary to sustain the body—hence their value to the poor. They are *indirectly* nutritious.

c. Specially, they diminish the quantity of carbonic acid given off from the lungs in a given time—(PROUT)—and that also of urea, phosphoric acid, and common salt in the urine. (JULIUS LEHMANN.) These are the chemical forms in which the lessening of the change of matter manifests itself. In the case of coffee it has been ascertained by experiment, that this lessening of the waste is due more to the empyreumatic oil than to the caffeine. The same is probably true also of tea.

d. The increased action of the heart, the trembling, the headache, and the peculiar intoxication and delirium which extreme indulgence in coffee sometimes produces, are mostly caused by the caffeine. On the other hand, the increased action of the kidneys, of the bowels, and of the perspiring vessels, and generally the increased activity of the whole system, are ascribed to the action of the oil. That Chinese tea has an astringent or costive effect upon the bowels, may arise either from the empyreumatic oil of tea not acting in the same way as that of coffee, or from the larger proportion of the astringent tannic acid which tea contains being able to counteract the effect of the oil. That there is a specific difference in the action of the empyreumatic oils of tea and maté, compared with that of coffee, is further probable from the remarkably intoxicating effect which both the Chinese and the Paraguay leaves possess when newly gathered and roasted for use.

Of course the general effect of these beverages upon the system is the combined result of the simultaneous action of all their constituent ingredients. But possessing the two characteristic influences of retarding the change of matter, and of increasing at the same time the activity of the nervous life, they cannot, according to our present knowledge, be replaced by the strongest soups or flesh teas, or by any other infusions or decoctions which merely supply the ordinary kinds of nourishment in more or less diluted and digestible forms.

In some countries it is the custom to heighten the natural flavour of roasted coffee by the addition of spices. Thus M. de Saulcy, in his recent tour round the Dead Sea, found the Bedouins in the country of ancient Moab drinking coffee, of which he says that it was "an absolute decoction of cloves." * On the Continent, and in North and South

* *Journey round the Dead Sea*, vol. i. p. 313.

America, vanilla is said to be employed largely for flavouring coffee as well as chocolate. To the other more natural influences of coffee these spices add a stimulating effect, which appears to expend itself chiefly upon the animal propensities.

A perusal of the history of these beverages leaves lingering in our minds some interesting general facts, which are suggestive of many thoughts.

The first is, the vast extent to which the materials for these beverages are cultivated and used, and the important place they occupy among what may be called the artificial necessities of life. Our data for forming correct calculations as to the quantity of each beverage which is grown and consumed are very defective, but we may guess them at about—

Chinese Tea, . .	2240	millions	of pounds.
Maté, . . .	20	"	"
Coffee, . . .	600	"	"
Chicory, . . .	80	"	"
Cocoa, . . .	100	"	"

forming an aggregate of nearly 3000 millions of pounds of the raw materials consumed annually in the preparation of the beverages we infuse.

Nor is the number of people to whom these warm beverages have become necessaries of life less surprising. Thus—

	Is consumed in	By about		
Chinese tea, .	China, Russia, Tartary, England, Holland, and North America.	500	millions	of men.
Maté or Paraguay tea,	Peru, Paraguay, Brazil, &c.	10	"	"
Coffee-tea,	Sumatra, &c.	2	"	"
Coffee-bean,	Arabia, Ceylon, Jamaica, Germany, France,	100	"	"
Chicory,	Germany, Belgium, France, England,	40	"	"
Cocoa,	Spain, Italy, France, Central America,	50	"	"

So that upon these four plants about three-fifths of the whole human race are dependent for one of their most useful and most harmless forms of indulgence.

A second point which strikes us in the history of these beverages—at least of the teas and coffees—is, that they have come more and more into use in Europe and America, as the intellectual activity which distinguishes the leading nations of modern times has developed itself. The kind of ordinary food upon which the consumers of these beverages usually live no doubt modifies the influence they exercise upon the system. It is even probable that the nature of this food is one of the causes which determine the preference given to tea or to coffee by the different European nations. And, reasoning from this probability, we might say that there is too much of mere vulgar nutrition in cocoa to allow it to influence the nervous or intellectual life to an equal degree with tea and coffee; and in this we might find a reason for the less prominent intellectual position which has been occupied by Spain and Italy, since cocoa has become an article of such universal consumption amongst them.

A third striking fact is, that the poorest and humblest among us, who has his own little earnings to spend, devotes a small part of it to the purchase of tea or coffee. He can barely buy bread and milk, or potatoes and salt, yet the cup of tea or coffee is preferred to the extra potato or the somewhat larger loaf. And if thereby his stomach is less filled, his hunger is equally stayed, and his comfort, both bodily and mental, wonderfully increased. He will probably live as long under the one regimen as the other; and while he does live, he will both be less miserable in mind, and will show more blood and spirit in the face of difficulties, than if he had denied himself his trifling indulgence. Besides the mere brickwork and marble, so to speak, by which the human body is built up and sustained, there are rarer forms of matter, as these chapters have shown, upon which the life of the body and the comfort of animal existence most essentially depend. This truth is not unworthy the consideration of those to whom the arrangement of the dietaries of our prisons, and other public institutions, has been intrusted. So many ounces of gluten, and so many of starch and fat, are assigned by these food-providers as an ample allowance for everyday use.* From these dietaries, except for the infirm and the invalid, tea and coffee are for the most part excluded. And in this they follow the counsel of those who have hitherto been regarded as chief authorities on the chemistry of nutrition. But it is worthy of trial whether the lessening of the general bodily waste, which would follow the consumption of a daily allowance of coffee, would not cause a saving of gluten and starch equal to the cost of the coffee;—and should this not prove the case, whether the increased comfort and happiness of the inmates, and the greater consequent facility of management, would not make up for the difference, if any. The inquiry is an interesting one in physiological economies, and it is not undeserving of the serious attention of those benevolent minds which, in so many parts of our Islands, have found in the prisons and houses of correction their most favourite fields of exertion.

I might add, as a stimulus to such experiments, the evident craving for some such indulgence as a kind of natural necessity, which is manifested in the almost universal practice among every people not absolutely savage, of preparing and drinking beverages of this sort. If there be in the human constitution this innocent craving, it cannot be misplaced humanity to minister to it, even in the case of the

* See the Author's *Elements of Agricultural Chemistry and Geology*, sixth edition, p. 394.

depraved and convicted. Where reformation is aimed at, the moral sense will be found most accessible where the mind is maintained in most healthy activity, and where the general comfort of the whole system is most effectually promoted.

CHAPTER X.

THE SWEETS WE EXTRACT.

THE GRAPE AND CANE SUGARS.

Mineral sweets.—Vegetable sweets.—Number of these known to modern nations.—The grape sugars; their sensible and chemical characters.—Honey sugars.—Trebizond honey.—Poisoning of Xenophon's soldiers.—Fruit sugars.—Starch or potato sugar, manufacture of.—Sugar from rags, from sawdust, and from Carrigeen, Ceylon, and Iceland mosses.—The cane sugars.—Spread of the sugar cane from Asia through Europe to America.—Varieties of the sugar cane.—Nutritive qualities of the raw cane juice.—Extensive consumption of it.—Composition of the sugar cane.—Manufacture of cane sugar.—Difficulties in the manufacture.—Great loss of sugar in consequence.—Improvements in the manufacture, and their effects on West Indian prosperity.—Total produce of cane sugar in the world.—Consumption of sugar in the United Kingdom.—Sensible and chemical characters of cane sugar.—Beet or European sugar.—Its importance on the continent of Europe.—Number and produce of the manufactories of France, Germany and Russia.—Composition of the sugar beet.—Difficulties in extracting the sugar.—Progress of the manufacture.—Its chemico-agricultural relations.—Palm or date sugars.—Quantity produced yearly.—Maple, or North American sugar.—Quantity produced in Canada, New England, and New York.—Mode of extraction.—Chemical changes in the maple sap.—Maize, or Mexican sugar; manufacture of, in the United States, and in France.—Sorghum sugar, the cane of the north.—Total quantity of sugar extracted for use.—Chemistry in its economical and social relations.

In common life, the sweets we extract are a constant accompaniment of the beverages we infuse. At least, as we use them in Europe and America, sugar is a usual addition to the infusions of tea, coffee, and cocoa.

Of substances which are sweet to the taste, the chemist is familiar with many which have no relation to the wants or usages of common life. Sugar of lead is a well-known poison, which derives its name from the sweetness of its taste. Silver in certain of its compounds * is equally sweet. A mineral earth called glucina (from γλυκυς, sweet), produces many compounds which have a sugary taste when first put into the mouth; and numerous other instances might be named. It is only those sweet substances, however, which exist in or are extracted from plants, that are directly connected with our modern comforts. These sweets not only accompany, on our tables, the beverages we infuse, but are the ingredients from which our brewers and distillers manufacture the liquors we ferment. They fall naturally to be considered, therefore, in this place.

Of these vegetable sweets, modern nations use many varieties. In such substances as luxuries of life, we are, indeed, far richer than any of the ancient nations. Thus, to the honey, grape, manna, and fruit sugars, which were the principal sweets of the ancient world, we now add the cane, maple, beet, maize, and palm sugars. We manufacture sugar also from potatoes and other substances rich in starch; from sea-weeds gathered by the shore; even from sawdust when an emergency arises; and we extract it from the milk of our domestic cattle. It has become to us, in consequence, almost a necessary of life. We consume it in millions of tons; we employ thousands of ships in transporting it. Millions of men spend their lives in cultivating the plants from which it is extracted, and the fiscal duties imposed upon it add largely to the revenue of nearly every established government. It may be said, therefore, to exercise a more direct and extended influence not only over the social comfort, but over the social condition of mankind, than any other production of the vegetable kingdom, with the exception, perhaps, of cotton alone.

* One called the *hyposulphite of silver*, for example, is very sweet.

The numerous varieties of useful sugars with which we are acquainted, may be arranged under four main kinds or heads. These are the grape sugars, the cane sugars, the manna sugars, and milk or animal sugar. I shall treat of each in its order.

I. The Grape Sugars include, as varieties, the sugar of the grape, the sugars of honey, the sugar of fruits, and potato or starch sugar.

1°. *Grape sugar.*—When the ripe grape is dried in the air, it forms the well-known raisin of commerce. When this raisin is opened, numerous whitish crystalline brittle granules are seen within it, which are sweet to the taste. These consist of what is called grape sugar, and they are the source of the sweetness both of the grape and the raisin. It dissolves readily in water, and if yeast be added to the solution, soon enters into fermentation.

The results of this fermentation are, first, a spirituous liquor resembling weak wine, and afterwards, as the fermentation proceeds, an acid liquor, like sour wine or vinegar.

In Syria, a sweet preparation is made from the juice of the grape. It consists chiefly of grape sugar, and is exported to Egypt under the name of *dips* or *dibs.**

2°. *Honey sugars.*—The bee has been long known and admired for its industry, and the honey it collects indulged in as a luxury. This honey is formed, or naturally deposited, in the nectaries of flowers, and is extracted from them by the working bees. They deposit it in their crop or honey-bag, which is an expansion of the gullet (œsophagus), and from this receptacle they disgorge it again when they return to the hive. In the interval, it is probably somewhat

* In Genesis, xliii. 11, this word is translated *honey*, though the sweet of the grape is probably meant. *Dibs* is also the word used for Samson's honey (Judges, xiv. 8), though *Assal* is also the word now employed in Syria and Egypt to denote the honey of the bee.

altered by admixture with the liquids which are secreted in the mouth and crop of the insect—so that the honey we extract from the hive may not be exactly in the same chemical condition as when it was sucked up from the flowers by the laborious bee.

When liquid honey is allowed to stand for a length of time, it gradually thickens and consolidates. By pressure in a linen bag, it may then be separated into a white solid sugar, consisting of minute crystals, which remain in the bag, and a thick semi-fluid syrup which flows through it. In old honey the proportion of syrup is often small, the sugar of the syrup gradually crystallising in greater quantity. Both the solid and liquid sugars have the same general properties. They are both equally sweet; both have the same chemical composition, and both begin to ferment when water and a little yeast are added to them. The solid sugar of honey is identical with the sugar of the grape. The liquid sugar differs from the solid chiefly in refusing to crystallise, and in containing an admixture of colouring and odoriferous substances produced by the flowers from which the bee has extracted it.

To these foreign substances honey owes the varied colours, flavours, and fragrances, which in different countries and districts it is known to possess, and for which it is often highly prized. Hence the estimation in which the honey of Mount Ida, in Crete, has been always held. Hence also the perfume of the Narbonne honey, of the honey of Chamouni, and of our own high moorland honey when the heather is in bloom. Sometimes these foreign substances possess narcotic or other dangerous qualities, as is the case with the Trebizond honey, which causes headache, vomiting, and even a kind of intoxication in those who eat it. This quality it derives from the flowers of a species of rhododendron (*Azalea* pontica), from which the honey is partly extracted. It was probably this kind of honey which poisoned the soldiers of Xenophon, as described by him in the retreat of the Ten Thousand.*

3°. *Fruit sugars.*—Many of our fruits pass, in the course of ripening, from a sour to a sweet state. The apple, the pear, the plum, the peach, the gooseberry, the currant, the cherry, &c., are of this kind. Most of them, even when fully ripe, are still a little acid; the mixture of sweet and sour in their juices, adding to their agreeable and refreshing qualities. All such fruits, as a general rule, contain, and owe their sweetness to, grape sugar. From many of them this sugar can be readily extracted for use; but, in general, it is more economical and agreeable to employ it in the form of dried and preserved fruits, or to make wine of it, as we do of that which exists in the grape, the gooseberry, the apple, and the pear.

4°. *Potato or starch sugar.*—It is a property of starch of all kinds to be insoluble in cold water, but to dissolve readily in boiling water, and to thicken into a jelly or paste as it cools. Even a lengthened boiling in water, however, produces little further change upon it. But if a small quantity of sulphuric acid (oil of vitriol) be added to the water in which it is boiled, the solution gradually acquires a sweet

* The effects of this honey upon his soldiers are thus described by Xenophon—"And there was there (in a village near Trebizond) a number of bee-hives; and as many of the soldiers as ate of the honey-combs became senseless, and were seized with vomiting and diarrhœa; and not one of them could stand erect. Those who had swallowed but little, looked very like drunk men; those who ate much were like madmen; and some lay as if they were dying. And thus they lay in such numbers, as on a field of battle after a defeat. And the consternation was great. Yet no one was found to have died; all recovered their senses about the same hour on the following day. And on the third or fourth day thereafter, they rose up as if they had suffered from the drinking of poison."—XENOPHON, *Anabasis*, book iv. chap. 8. Τα δε σμηνη, &c.

Auguste St. Hilare, while travelling in the Brazils, experienced symptoms of poisoning after having eaten of honey extracted by a native bee from a plant belonging, to the poisonous family of *Apocynaceæ*, or dogbanes.

taste, and ultimately the whole of the starch is converted into grape or honey sugar. A pound of acid diluted with a hundred pounds of water, and employed in this way, will convert into sugar a great many pounds of potato, wheaten, or sago starch. If the acid be then separated by lime, and the liquor boiled down, either a rich syrup or a solid sugar may be obtained. Or, instead of sulphuric acid, we may mix with the water, 12 or 15 lbs. of malt for every 100 lbs. of starch; heat for three hours to 160° or 170° Fahr., and then filter and evaporate the syrup. Sugar thus prepared from starch has the same sweetness, chemical composition, and general properties as that of the grape. It does not always crystallise readily, however, and in this respect has more resemblance to the liquid sugar of honey than to the solid sugar of the dried grape. It is used for ordinary sweetening purposes, for adulterating cane sugar, and for the manufacture of spirituous liquors. On the continent of Europe it is largely prepared for all these uses. The syrup is extensively employed by the French confectioners, and brandy distilled from it is very generally drunk in northern Europe. The manufacture of starch sugar is illegal in this country.

Instead of starch, woody fibre may be employed for the manufacture of this kind of sugar. Paper, raw cotton and flax, cotton and linen rags, and even saw-dust, may be transformed into sugar by digestion in diluted sulphuric acid. The operation is only a little slower, and therefore requires more time. This is partly explained by the fact that the acid first changes the fibre into starch, and then the starch further into sugar.

It is known that many sea-weeds, when boiled in water, yield a jelly which is wholesome, nutritious, and more or less agreeable to the palate. Among these are the well-known Carrigeen moss (*Chondrus crispus* and *mamillosus*), which is collected in large quantities on the west coast of Ireland, and the Ceylon moss (*Plocaria Candida*), which is exported from the islands of the Indian Archipelago to the markets of China. The jelly yielded by these seaweeds, as well as by the Iceland and other land mosses, is in like manner converted into grape-sugar, when digested with diluted sulphuric acid.

The number of vegetable substances, therefore, which by means of this acid can be transformed into the sugar of honey and fruits, is very great. Starch, however, is the only one to which the process has hitherto been applied with a profit. The way in which these singular transformations of matter are brought about, will be illustrated at the close of the succeeding chapter.

5°. *Elderberry sugar.*—In the berries of the elder tree (*Sorbus aucuparia*), a peculiar species of sugar has recently been discovered, to which M. Pelouze has given the name of *sorbine.* In the degree of sweetness it possesses, and in chemical composition, it agrees with grape sugar; but it differs from it in its other properties, and in its crystalline form. As yet, however, this variety of sugar is of no economical value.

II. The Cane Sugars.—The plants or fruits which possess distinctly acids or sour juices, yield grape sugar. Those which have little acid in their saps, contain for the most part cane sugar. The chemical reason for this is, that, by the action of acid substances, cane sugar is gradually transformed into grape sugar, even in the interior of the growing plant. The principal varieties of cane sugar known in commerce, are the cane sugar properly so called, beet sugar, palm or date sugar, maple sugar, and maize sugar.

1°. *Sugar cane or Chinese Sugar.*—The sugar cane (fig. 36) is the chief source of the sugar of commerce. About eleven-twelfths of all the sugar extracted for use is obtained from this plant. Though almost unknown to the Greeks and Romans, and now cultivated most extensively in America, it is a native of the Old World. It was familiar in the East in most remote times, and appears to have been cultivated in China and the South Sea Islands long before the period of authentic history. Through Sicily and Spain it reached the Canary Islands, thence was transplanted to St. Domingo by the Spaniards in 1520, and from this island it has gradually spread over the West Indies and the tropical regions of the American continent. It flourishes best where the mean temperature is from 75° to 77° Fahr.; but it thrives, and can be economically cultivated where the mean temperature does not exceed 66° to 68° Fahr. Hence it is grown far beyond the tropics. And although the countries most productive in sugar, and which yield it at the least cost, lie for the most part within the torrid zone and at low elevations, —yet the sugar cane is profitably grown in some parts of the south of Europe; on the table-land of Nepaul, in India, at a height of 4500 feet, and on the plains of Mexico, as high as 4000 to 6000 feet above the level of the sea. It rarely ripens its seed, however, even in the most propitious localities. Young plants are raised, therefore, from portions of the stem planted for the purpose; and when cultivated for

Fig. 86.

Saccharum officinarum—
The Sugar Cane.
Scale. 1 inch to 4 feet.

sugar, they are rarely allowed to come to flower as is represented in fig. 36.

There are many varieties of the sugar cane, as there are of nearly all long-cultivated plants. In general, the varieties most common in each country and district are best adapted to the local climate and to the soils in which they grow. Those which yield the sweetest juice, and in the greatest abundance, if otherwise suited to the climate, are the most esteemed. In Louisiana, five different varieties are cultivated, one of the most elegant of which is represented in the annexed drawing, (fig. 37.) In each locality that variety is selected by the planter which he finds to give, on the whole, the most sure and profitable crop.* And so in our West India colonies the Tahati cane was introduced as a new variety, because in the same time, and from the same extent of land, it yielded one fourth more juice than the common varieties, while it produced also a larger and more solid growth of wood to be used as fuel.†

Fig. 37.

Striped Cane of Louisiana.

In Europe and

* American Patent Office Report, 1848. P. 281.
† MEYEN, *Geog. of Plants*, p. 382.

most northern countries, cane sugar is only an article of luxury, though one with which many would now find it difficult to dispense. In many tropical regions, however, the sugar cane forms a staple part of the ordinary food. The ripe stalk of the plant is chewed and sucked after being made soft by boring it, and almost incredible quantities are consumed in this way. Large ship-loads of raw sugar cane are daily brought to the markets of Manilla and Rio Janeiro; and it is plentiful in the market of New Orleans. In the Sandwich and many other islands of the Pacific, every child has a piece of sugar cane in its hand; while in our own sugar colonies the negroes become fat in crop time on the abundant juice of the ripening cane. This mode of using the cane is, no doubt, the most ancient of all, and was well known to the Roman writers. Lucan (book iii. 237) speaks of the eaters of the cane as—

"Quique bibunt tenerâ dulces ab arundine succos."

—"And those who drink sweet juices from the tender reed."

This nutritive property of the raw juice of the sugar cane arises from the circumstance that it contains, besides the sugar to which its sweetness is owing, a considerable proportion of gluten, as well as of those necessary mineral sub stances which are present in all our staple forms of vegetable food. It is thus itself a true food,* capable of sustaining animal life and strength without the addition of other forms of nourishment. This is not the case with the sugar of commerce, which, though it in a certain sense helps to nourish us, is unable of itself to sustain animal life.

The juice of the sugar cane varies in composition and richness with the variety of cane, the nature of the soil, the mode of cultivation, and the dryness of the season. Its

* See THE BREAD WE EAT.

average composition in sugar plantations, when the canes are fully ripe, is about—

Sugar,	.	18 to 22
Water and gluten	.	71
Woody fibre,	.	10
Saline matter	.	1
		100

The richness in sugar varies with many circumstances, and especially with what is called the ripeness of the cane. For it is a curious circumstance in the chemical history of this plant that the sap sweetens only to a certain distance up the stem; the upper somewhat green part, which is still growing, yielding abundance of sap, but comparatively little sugar. One reason of this probably is, that as fast as the sugar ascends with the sap, it is converted into woody matter, which is built in to the substance of the growing stem and leaves. In consequence of this want of sweetness, the upper part of the cane is cut off, and only the under ripe part employed in the manufacture of sugar. In Louisiana, where the canes rarely ripen so completely as in the West Indies, the proportion of sugar contained in the juice is set down as low as 12 to 14 per cent.*

For the extraction of the sugar, the canes are cut with a large knife, the labourer proceeding between the rows (fig. 38). The leaves and tops are then chopped off and left in field, while the under ripe part is carried to the mill. These ripe canes are passed between heavy iron crushing-rollers, which squeeze out the juice. This juice is run into large vessels, where it is clarified by the addition of lime and other applications. The action of this lime is twofold. It removes or neutralises the acid which rapidly forms in the fresh juice, and at the same time combines with the gluten

* Patent Office Report, 1844.

of the juice, and carries it to the bottom. This gluten acts as a natural ferment, causing the sugar to run to acid. Its speedy removal, therefore, is essential to the extraction of the sugar. After being clarified in this way, and sometimes filtered, the juice is boiled rapidly down, is then run into wooden vessels to cool and crystallise, and, finally, when crystallised, is put into perforated casks to drain. What remains in these casks is Muscovado or raw sugar; the drainings are well known by the name of molasses.

Fig. 88.

Cane plantation in Louisiana.

Simple as this process is in description, it is attended with many difficulties in practice. It is difficult to sqeeze the whole of the juice out of the cane—it is difficult to clarify the juice with sufficient rapidity to prevent it from fermenting, and so completely as to render skimming unnecessary during the boiling—it is difficult to boil it down rapidly without burning or blackening, and thus producing much uncrystallisable molasses—and it is difficult afterwards to collect and profitably employ the whole of the molasses thus produced. The difficulties, though none of them insurmountable, have hitherto proved so formidable in practice, that, of the 18 per cent. of sugar contained in the average cane-juice of our West India Islands, not more than 6 per cent., or one-third of the whole, is usually sent to market in the state of crystallised sugar! The great loss which thus appears to take place is thus accounted for—

First,—Of the 90 per cent. of sweet juice which the cane contains, only 50 to 60 per cent. are usually expressed. Thus one-third of the sugar is left in the megass, or squeezed cane, which is used for fuel—(KERR.)

Second,—Of the sugar in the juice, one-fifth or more is lost by imperfect clarifying, and in the skimmings removed during the boiling—(SHEIR.)

Third,—Then of the juice when boiled down to the crystallising point and set to cool, only from one-half to two-thirds crystallises: the rest drains off as molasses. Thus of the whole sugar of the ripe cane—

One-third is left in the megass, . .	6	per cent.
One-third of the remainder in the skimmings,	2½	"
One-third to one-half of the second remainder in the molasses, ...	3	"
In the Muscovado sent to market there are	6½	"
	18	

The molasses and skimmings are fermented and distilled for rum. But even of the molasses much is lost, the drainage from the raw sugar of the West Indies, while at sea, is stated at 15 per cent., and afterwards, in the docks, at 2 per cent. And further, the leakage of the molasses itself, which is shipped as such, is 20 per cent.; so that of the un-uncystallisable part of the sugar, also, there is a large waste. In the interior of Java, where fuel is scarce, the molasses is worthless, and is sent down the rivers in large quantities; but in the West Indies it has everywhere a market value, and may be distilled with a profit.

The sugar manufacture, therefore, of our West India

colonies, appears *as a whole* to be in a most unsatisfactory condition. Neither mechanical nor chemical means have been applied to it as they have been to the sugar manufacture of Europe; and it is not at all surprising that pecuniary difficulties should of late years have gathered round the unimproving planters. The same skill which now extracts 7 per cent. of refined sugar from the more difficult beet, might easily extract 10 or 12 from the sugar cane. Were this result generally attained, the same weight of canes which is now grown in the West Indies, and which yields less than half the quantity of crystallised sugar actually consumed in the United Kingdom, would alone produce enough to supply the entire present home consumption.

The means by which this better result is to be attained are, the use of improved crushing rollers, by which 70 and even 75 per cent. of juice can be forced from the canes—of better modes of clarifying, which chemical research has recently discovered—of charcoal filters before boiling, which render skimming unnecessary—of steam and vacuum boilers, by which burning is prevented, and rapid concentration effected—of centrifugal drainers to dry the sugar speedily and save the molasses—and of coal or wood as fuel where the crushed cane is insufficient for the purpose. By the use of such improvements, planters in Java, in Cuba, and, I believe, here and there in our own colonies, are now extracting and sending to market 10 to 12 per cent. of raw sugar from the 100 lb. of canes! Why should our own enterprising West India proprietors spend their time in vain regrets and longings for the past, instead of earnestly availing themselves of those scientific means of bettering themselves which are waiting to be employed, and which are ready to develop themselves to meet every new emergency? It is not the readier or cheaper supply of labour which gives the Dutch planter of Java, or the Spanish planter of Cuba, 10 per cent. of marketable sugar, but better machinery and more refined chemical applications. And these are surely as much within the reach of British subjects as of any other people on the face of the earth.

The total quantity of sugar extracted from the sugar cane over the whole globe, has been estimated by Stolle at 4527 millions of pounds. Of this the largest proportion is yielded by the British East and West Indies. The consumption in the United Kingdom amounts at present to about two-elevenths of the enormous quantity above stated. In 1853 our home consumption amounted to 818 millions of pounds of raw sugar. This is equal to 28 lb. of sugar per head of the population, and the quantity is rapidly increasing. How wonderful a change in the tastes and habits of the people does this imply since the year 1700, when the quantity consumed in England was only 22 millions of pounds! And the consumption per head in Great Britain is considerably more than the above 28 lb., because the average consumption per head in Ireland, of which no separate account has been kept since 1826, is not more than one-third of the British consumption.

An acre of land in the West Indies yields, according to the present mode of extraction, from 1 to 3, or even 4 tons of sugar, and for each ton of sugar about 70 gallons, or 1400 lb. of marketable molasses. At an average of 3 tons an acre of sugar and molasses, it requires upwards of 130 thousand acres of rich land to produce the sugar yearly consumed in the British Islands!

The cane sugars are popularly distinguished from the grape sugars by greater sweetness or sweetening power. This is said to be greater in the proportion of five to three.*

* The sense by which we appreciate the sweetness of bodies is liable to singular modifications. Thus the leaves of the *Gymnema sylvestre*—a plant of Northern India—when it is chewed, takes away the power of tasting sugar for twenty-four hours, without otherwise injuring the general sense of taste.

They also dissolve more readily in water. One pound of cold water dissolves 3 lb. of cane, but only 1 lb. of grape sugar. The solution is also thicker and more syrupy, less liable to change or run to acid, crystallises more readily, and gives a harder candy. These superior economical properties sufficiently account for the preference so universally given to this species of vegetable sweet.

Chemically the cane differs from the grape sugars, in containing less of the elements of water, in being charred or blackened by strong sulphuric acid (oil of vitriol), and in not readily throwing down the red oxide of copper from solutions of blue vitriol (sulphate of copper). By the action of diluted acids cane sugar is converted into grape sugar, and hence the reason why, as I have already said, cane sugar is rarely found in plants which have acid juices, and why the souring of the cane juice changes a portion of its crystallisable sugar into uncrystallisable syrup or molasses.

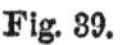
Fig. 39.

Sugar Beet. Scale, half an inch to a foot.

2°. *Beet-root or European sugar.*—The root of the beet, and especially of the variety called the sugar-beet (fig. 39), contains often as much as a tenth part of its weight of sugar. By squeezing out the juice, as in the case of the sugar cane, or by dissolving out the sugar from the sliced root and boiling down the solution, the raw sugar is obtained. In this state the sugar possesses a peculiar, unpleasant flavour, derived from the beet-root; but when refined, it is scarcely distinguishable in any respect from that of the sugar cane.

The manufacture of this sugar is one of great and growing importance, especially in France, Belgium, Germany, and Russia. Its history also illustrates in a very striking way how chemical skill may overcome, as it were, the perversities of climate, and establish, upon an artificial basis, an important national interest, which shall successfully compete in the markets of the world with the most favoured natural productions of the choicest regions of the globe.

As early as 1747, Margraaf, in Berlin, drew attention to the large quantity of sugar contained in the beet, and recommended its cultivation for the manufacture of sugar. Fifty years later the attempt was made in Silesia, under royal patronage; but as only two or three per cent. of crystallised sugar could be extracted, the work failed and was abandoned. Later, again, the continental system of Napoleon I. which raised the price of sugar to five shillings (six francs) a pound, and especially the offer of a prize of a million of francs for the successful manufacture of sugar from plants of home growth, stimulated to new trials both in Germany and France. New methods, new skill, new machinery, and the results of later chemical research, were all applied, and with the aid of high duties on foreign sugar, the manufacture struggled on through a period of very sickly infancy. In Germany fewer improvements were introduced, so that the new manufactories erected in that country, during the reign of Napoleon were one after another given up; but in France they became so firmly established, that even after the cessation of the continental system few of them were abandoned. A more complete extraction of the sap, a quicker and easier method of clarifying and filtering it, and the use of steam to boil it down, enabled the French maker to extract 4 to 5 per cent. of refined sugar from the 100 lbs. of beet, and thus to conduct his operations with a profit. In this improved condition the manufacture, after a struggle of twenty years, returned again towards the north, and spread not only over Belgium and the different states of Germany, but over Poland, and into the very heart of Russia. At the present time, not less than 362 millions of pounds of beet sugar are manufactured on the continent of Europe. This is equal to about $7\frac{1}{3}$ per cent. of all the sugar consumed in the world. The proportion extracted in the different countries named is nearly as follows :—

	Number of Manufactories.	Average produce of each Manufactory.	Total Produce of the country.
Russia, . . .	360	200,000 lbs.	70,000,000 lbs.
France, . . .	334	440,000 "	150,000,000 "
German Customs Union,	237	560,000 "	130,000,000 "
Belgium, . . .	30	400,000 "	12,000,000 "

There are, besides, some manufactories in Austria which produce on an average 160,000 pounds of sugar each.

The extraction of sugar from the beet has lately been attempted in Ireland, and, as I am informed, with some measure of success. Little is publicly known, however, of the proceedings of the company by which the attempt has been made.

The average composition of the root of the sugar beet of France, Belgium, and the Rhenish provinces, is nearly as follows :—

Sugar, ...	$10\frac{1}{2}$
Gluten, ...	3
Fibre, &c., ...	5
Water, ...	$81\frac{1}{2}$
	100

But this proportion of sugar varies very much. Thus it is greater,—

a In small beets than in large.

b In some varieties, as in the white Schleswick pear-shaped beet, and in a spindle-shaped, red-skinned, white-fleshed variety, both much cultivated in Germany.

c In dry climates, and especially where the climate is dry after the roots have begun to swell.

d In light potato or barley than heavy soils.

e In the part under than above ground.

f When manure has not been directly applied to the crop.

These facts show how much practical agriculture has to do with the success of this important manufacture. The difference of climate, soil, and mode of culture have so much effect, that, while the beets of Lille, a southern centre of the manufacture, do not average more than 10 to 12 per cent. of sugar, those of Magdeburg, a more northern centre, contain from 12 to 14 per cent. Under certain very favourable conditions, as much as 18 per cent. of sugar has been found in the beet of North Germany. The proportion of sugar is so much less in the part that grows above ground, that it is not unfrequently cut off and fed to cattle. This reminds us of the want of sweetness in the upper part of the sugar cane, (p. 251), and the reason is probably the same in both cases, that the sugar is in these parts transformed into woody matter.

The average proportion of sugar extracted in Belgium and France, is 6 lb. from every hundred of fresh root. In some well-conducted manufactories, it is said to reach 7, and even 7½lb. from the hundred. In Germany, the average yield is at present 7 or more; and improvements now on trial are expected to raise it to 8 lb. from the hundred.

The mode of extraction is very simple. In France and Belgium the root is ground to a pulp between saw-toothed rollers, a small stream of water trickling over the teeth to keep them clean. This pulp is put into bags, and submitted to strong pressure, by which the juice is squeezed out, while the solid matter remains in the form of a dry cake. This juice is treated with lime, heated, filtered, boiled down by steam to the crystallising point, and then, as in the case of cane sugar, cooled and drained from the molasses. From the beet, the molasses thus obtained is colourless, but it has a disagreeable taste, and cannot, therefore, like cane molasses, be directly employed for any sweetening purpose. The raw sugar has also an unpleasant taste, and is in consequence refined, for the most part, before it is brought to market.

In Germany, it is more usual to slice the beet, and to wash out the sugar with hot water, treating the solution afterwards as above described. The happy discovery of Melsens, of Brussels, that sulphurous acid * has the property of arresting fermentation in sweet juices, has been of much service in making this German method of extraction available.

It is interesting to remark how new improvements in this manufacture constantly make known new chemical difficulties, and present new chemical and agricultural problems to be solved. The first great difficulty was, to prevent the fermentation of the juice, the production of acid, and the simultaneous waste of sugar and conversion of a part of it into uncrystallisable syrup. The second was, to boil it down so as to prevent burning, and the production of uncrystallisable molasses. The former has been overcome by various chemical means, and the latter by the use of steam. But as the yield of sugar approached to 7 per cent., it was found

* Sulphurous acid is the name given by chemists to the strong-smelling fumes given off by burning sulphur. In one proportion, it forms with lime *sulphite of lime;* in twice this proportion it forms *bi*-sulphite (*bis* twice). This bisulphite is soluble in water, and a little of the solution added to the weak sugary liquors prevents them from fermenting.

that certain syrups remained behind, which, though they certainly contained cane sugar, refused stubbornly to crystallise; and the reason of this was traced to the presence of saline matter, chiefly common salt, in the sap. This salt forms a compound with the sugar, and prevents it from crystallising. And so powerful is this influence, that 1 per cent. of salt in the sap will render 3 per cent. of the sugar uncrystallisable. To overome this difficulty, new chemical inquiries were necessary. As results of these inquiries, it was ascertained—

First, That the proportion of sugar was larger, and of salt less, in beets not weighing more than five pounds. The first practical step, therefore, was, that the sugar manufacturers announced to the cultivators who raised the beet, that in future they would give a less price for roots weighing more than five pounds.

Next, That a crop raised by means of the direct application of manure, contained more salt, and gave more uncrystallisable syrup, than when raised without direct manuring. A larger price, therefore, was offered for roots grown upon land which had been manured during the previous winter; a higher still for such as were raised after a manured crop of corn; and a still higher when, after the manuring, two crops of corn were taken before the beet was sown.

Thus, the difficulty was lessened by chemico-agricultural means; and though the crop was less in weight to the farmer, the higher price he obtained in some degree made up the difference.

In France and Belgium, the crops gathered average 14 or 15 tons an acre, while about Magdeburg they do not exceed 10 or 12 tons. But the latter are richer in sugar, and poorer in salts, in proportion. A method is now under trial in France for separating the sugar from the salts by a purely chemical operation. When this is effected, the crops may be forced by manure as our turnip crops are, and a larger yield obtained without fear of injuring the sugar extractor by a superabundance of salts.*

One other point in this history is very interesting, as illustrative of the way in which a tax upon manufacturing industry may be made actually to promote, instead of retarding its advancement! The tax on beet sugar within the bounds of the German Customs Union (Zollverein), is levied, not on the sugar actually produced, but upon the weight of raw beets employed by the manufacturer. It is assumed that the roots will yield 5 per cent., or one-twentieth of their weight of sugar; and then upon every 20 cwt. of roots a tax of two dollars is imposed. According to the assumed yield of sugar, this is equal to a tax of two dollars on every hundredweight of sugar. But in reality it is much less. By the improved methods, one of sugar can now be extracted from about fourteen of the root; and the more he can extract, the less duty in proportion the manufacturer pays. Thus he is continually stimulated to improve his methods. The absolute gain which he derives from an increased produce per cent., is enchanced by the peculiar satisfaction which arises from the consciousness that every additional pound he extracts is duty free.

And the profit he thus makes is at the same time a source

* It is stated, also, that Mr. Hertz, a partner in a large sugar manufactory near Heidelburg, has made improvements by means of which, among other advantages, he can extract from his beet 9 per cent. of pure sugar.

1°. He dries the roots whole in ovens, and thus can keep them all the year round and work them when most convenient.

2°. He washes out the Sugar in vacuo. This excludes the air, prevents fermentation, more fully extracts the sugar, enables him to work in warm as well as in cold, weather, and thus, in the course of the year, to work up three times the material with the same plant.

There may be some exaggeration in the alleged results of these methods, but the idea of extracting the sugar in vacuo is certainly good, if it can be economically effected, and generally applied.

of gain to others. It is the character of all scientific progress, that an advanced step taken in one country is at once a signal for similar steps in other countries, and an assurance that they will by-and-by be taken. Thus the improvements which arise out of the fiscal regulations of the German Zollverein are gradually introduced into the boiling-houses of Cuba, and, more slowly perhaps, yet certainly in the end, will render more perfect and profitable the planting operations of our own West India colonies.

3°. *Palm or date sugar*, or Jaggery.—Most trees of the palm tribe, when their top-shoot, or spadix as it is called, is wounded, yield a copious supply of sweet juice. When boiled down, this juice gives a brownish raw sugar, known in India by the name of jaggery. The date palm (*Phœnix dactylifera*, p. 94) affords this juice and sugar. The gommuti palm (*Saguerus saccharifer*), fig. 40, is still more productive, and, in the Moluccas and Philippines, yields much sugar. The sap of the cocoanut tree is boiled down in the South Sea Islands till it has the consistence of a brown syrup, resembling very much the molasses which drains from raw cane-sugar; but the wild date-palm (*Phœnix sylvestris*) is the largest known sugar-producer. From this tree it is said that 60,000 tons,* or 130 million pounds, are yearly extracted. Of this quantity, 5000 tons, or 11 mil-

Fig. 40.

Saguerus saccharifer—The Gommuti Palm.
Scale, 1 inch to 20 fee.

* Archer's *Popular Economic Botany*, p. 140.

lion pounds, are extracted in Bengal alone. Indeed, the chief production as well as consumption of this date sugar is in India. A small proportion of it is imported into this country, sometimes under its true name of jaggery, but often, also, under that of cane sugar.

This palm-sugar, indeed, from whichever of the trees it is extracted, is exactly the same species of sugar as is yielded by the sugar cane. It differs chiefly in the flavour of the molasses which drains from and colours the raw sugar. When refined, it cannot be distinguished from refined West India sugar. The flavour of the molasses is not unpleasant, so that it is readily eaten by the natives of the various tropical regions in which the palm trees grow.

Fig. 41.

Acer saccharinum—The Sugar Maple.
Scale, 1 inch to 30 feet.
Leaf, 1 inch to 5 inches.

The total known produce of palm sugar is estimated at 220 million pounds. This is bout one-twenty-fourth part of all the cane sugar extracted for useful purposes.

Other non-acid fruits, like the melon, the chestnut, and the cocoa-nut, contain cane sugar, but it is not extracted from them as an article of commerce.

4°. *Maple or North American sugar.*—The sugar maple (*Acer saccharinum*), fig. 41, grows abundantly in the northern parts of New England, along the lakes and in the British provinces of North America. The four States of New Hampshire, Vermont, New York, and Michigan produce together upwards of 20 million pounds, and the Canadas together about 7 million pounds of maple sugar. The settlers generally, when they clear their virgin farms, reserve a few trees to make sugar for the use of their families; but, in many places, extensive natural forests of maple trees still cover fertile tracts of uncultivated country, and there the sugar is manufactured in large quantities. The average yield of each tree is estimated in Lower Canada at 1 lb. a-head; and the right of making the sugar is there rented out by the proprietor at one-fifth of the supposed produce, or one pound of sugar for every five trees. When the month of March arrives, the sugar-makers start for the forest, carrying with them a large pot, a few buckets and other utensils, their axes and a supply of food. They erect a shanty where the maple trees are most numerous, make incisions into as many as they can visit twice a-day for the purpose of collecting the sap, boil down this sap to the crystallising point, and pour it into oblong brick-shaped moulds, in which it solidifies. In this way, in the valley of the Chaudière, from 3000 to 5000 pounds of sugar are sometimes made during the season of two months by a single party of two or three men.

It is a singular circumstance in the chemical history of the sap of this tree, that the first which flows for some time after the incision is made, is clear, colourless, and without taste. After standing a day or two this sap becomes sweet; and a few days after the sap has begun to run, it flows sweet from the tree. The last sap which the tree yields is thick, and makes an inferior sugar. When boiled carefully in earthenware, or glazed pots, the clear sap gives at once a beautifully white sugar, and especially if it be drained in moulds and clayed, as is done with common loaf-sugar. In this pure white condition it is not to be distinguished from refined cane-sugar. It is identical with pure cane sugar in all its properties.

For domestic use it is generally preferred of a brown, and by many of a dark-brown colour, because of the rich maple flavour it possesses. This flavour, though peculiar, and therefore new to a stranger in North America, soon becomes very much relished. The brown sugar is an article of regular diet among the Lower Canadians. On fast days, bread and maple sugar, or maple honey, as the molasses of this sugar is called, are eaten in preference to fish. In spring, when plentiful, it sells as low as 3d. a-pound; in winter it rises sometimes as high as 6d.*

It is an interesting character of the maple juice, when boiled to the crystallising point, that the molasses which drains from it is agreeable to the taste, and is relished as a domestic luxury. In this respect it is superior even to the molasses of the sugar cane. Were beet root molasses eatable in a similar way, the manufacture of beet sugar would have fewer difficulties to overcome; and it would have been now both easier to conduct and more profitable in its results.

The total production of maple sugar has been estimated at 45 millions of pounds, or the one hundred and twenty-fifth part ($\frac{1}{125}$) of the whole quantity of cane sugar extracted for the use of man. The manufacture of maple sugar diminishes yearly in proportion as the native American forests are cut down.

5°. *Maize or Mexican sugar.*—The green stalks of maize or Indian corn contain a sweet juice, which, when boiled down, yields an agreeable variety of cane sugar. This sugar was known and extracted by the ancient Mexicans, and was in use among them prior to the Spanish invasion. For this reason I have distinguished it as Mexican sugar.

The manufacture of this sugar has been attempted of late years in the United States, and many persons have success-

* See the Author's *Notes on North America*, vol. i. p. 303.

fully extracted a sufficiency for their domestic consumption. It has not hitherto, however, been prepared in such quantity, or at such a price, as publicly to compete in the market with sugar from the cane; but there seems no reason why this branch of industry should not be successfully prosecuted, especially in those States of the North American Union which are known to be more eminently favourable to the growth of maize.

The extraction of sugar from this plant has also been attempted in southern Europe. The only existing manufactory of it with which I am acquainted is in the south of France, in the neighbourhood of Toulouse. It produces only about 20,000 lb. of sugar a-year. But that this small manufactory can be profitably conducted in a climate less favourable to maize, affords a strong presumption that, in the United States, the cultivation of the plant for its sugar may yet become an important branch of rural economy.

Fig. 42.

Sorghum vulgare—Durrha plant.

6°. *Sorghum sugar.*—In China, under the name of "sugar cane of the north," a species of sorghum is cultivated for the extraction of sugar. This plant is allied to the *Sorghum vulgare*, or dhurra plant, (fig. 42), of which a description has already been given.* This plant has recently been introduced into France, and experiments have been made upon it by Mons. Vilmorin. He states that it is capable of yielding, on an average, from an acre of land, 26,000 lb. of juice, containing from 10 to 13 per cent. of

* See The Bread we eat, p. 89.

sugar; and that this is more than the average yield of the sugar beet. It is alleged, however, that the plant is adapted to only a few parts of the south of France. More will no doubt be heard of this plant should further experiments confirm the favourable opinions already formed of it.

The total quantities of cane sugar of various kinds, which are extracted for human use, have been estimated as follows by Dr. Stolle;—

	Millions of pounds.	Percentage of the whole production.
Cane sugar, . .	4527	87.7
Beet sugar, . .	862	7.3
Palm sugar, . .	220	4.2
Maple sugar, . .	45	0.8
	5154	100

Wide differences exist among the quantities consumed per head in different countries—I instance only a few examples. Thus, the yearly consumption is, in

Russia, . . .	1¼	lb. per head.
Belgium, . . .	5	"
France, . . .	7½	"
United Kingdom, . .	28	"
Venezuela, . .	180!	"

With the peculiar circumstances which occasion so large a consumption in Venezuela I am unacquainted. Refined sugar is shipped to that country largely from Europe.

Before leaving this part of my subject, I may be permitted, in the interest of chemical science, to ask my reader to reflect—

1°. How important an interest, economical and social, the history of sugar extraction exhibits to us as depending directly upon chemical research and progress, and upon the diffusion and application of chemical knowledge.

2°. How largely successive applications of this branch

of knowledge have already benefited the manufacture of sugar, and aided in bringing this luxury within the reach of the poorer classes; and how much more benefit they promise still to confer.

3°. And especially how chemistry has earned the deserved gratitude of the European continent, by giving it an entirely new industry, and by making it independent of foreign countries for one of the most esteemed and now almost necessary luxuries of life:

It is not the fault of chemistry that our West India colonies have not equal cause to be grateful.

CHAPTER XI.

THE SWEETS WE EXTRACT.

THE MANNA AND MILK SUGARS.

Manna sugars; their sensible and chemical characters.—Manna of the ash; its composition and uses.—Occurrence of manna sugar in sea weeds.—Gum-tree manna.—Other mannas.—Oak, larch, and cedar mannas.—Persian manna.—The alhagi and tamarisk mannas.—The manna of the Scriptures; trees supposed to produce it.—The real manna not known.—Liquorice sugar.—Milk sugar.—Analogies in the composition of cane, grape, and milk sugar.—How the two former are produced from each other, from starch, and from humic acid.—What chemists understand by chemical reactions.—How a knowledge of these improves old and gives rise to new chemical arts.—Illustration in the manufacture of garancine, and the use of madder in dyeing.

III. The Manna Sugars form a third class of sugars which are distinguished from the grape and cane sugars by three principal characters. *First*, by their chemical composition; *second*, by their inferior sweetness; and *third*, by their not fermenting when mingled with yeast. Of this class, also, there are several varieties.

1°. *Manna of the ash.*—Two species of ash, the *Fraxinus ornus*, and the *F. rotundifolia*, yield this species of sugar. The European supply is chiefly derived from Sicily and Calabria. The *F. ornus*, a small tree of twenty to twenty-five feet high, is there cultivated in plantations for the purpose. In the months of July and August, when the production of leaves has ceased, the sap is drawn from the tree. For this purpose, cross cuts, about two inches long (fig. 43), are made in the stem, beginning at the lower part near the soil. These are repeated every day in warm weather, extending them perpendicularly upwards along the one side of the tree, leaving the other to be cut in the following year. The sap flows from these incisions, and is sometimes collected in vessels and sometimes allowed to harden on the outside of the tree. It is very rich in sugar, and speedily concretes in fine weather into the manna of commerce. The quality of the manna varies with the age of the tree, and with the part of the stem (lower or higher) from which it flows, and with the period of the season in which it is extracted. From the upper incisions, from trees of middle age, and in the height of the season when the sap flows most freely, the flake manna, most esteemed in England, is obtained in largest quantity.

Fig. 43.

Fraxinus ornus—The Manna Ash, and the mode of collecting the manna.

Manna—besides a variable proportion of gum, which in some varieties amounts to a third of its weight—contains two kinds of sugar. The larger proportion consists of a peculiar, colourless, beautifully crystalline sugar, to which the name of *mannite* is given. This forms from 30 to 60 per cent. of the whole manna, and is properly *the* manna sugar. Mixed with this there is from 5 to 10 per cent. of a sugar resembling that of the grape, and which ferments with yeast. Thus, the manna of commerce consists, on an average, of about—

	Per cent.
Manna sugar, or mannite,	40
Grape sugar, (?)	10
Gum, with some gluten and other matters,	40
Water,	10
	100

The large admixture of gum diminishes the sweetness of the manna, and renders it less useful as a substitute for cane sugar.

When newly extracted, manna is found to be nutritious as well as agreeable to the taste; and a considerable quantity of it is used as food, especially in Calabria. As it becomes old, however, it acquires a mild laxative quality, which unfits it for use as a part of the ordinary diet. This latter quality recommends it for use as a medicinal agent, for which purpose it is exported to various parts of Europe. The quantity yearly imported into Great Britain amounts to about 11,000 lbs., nearly all of which comes from Sicily.

This medicinal quality does not reside in the mannite or true sugar of manna, but in the other matters with which it is contaminated. By itself, in the pure or refined state, this sugar has no appreciable medicinal action, and were it abundant and cheap, might be employed for ordinary sweetening purposes. It is less sweet than cane sugar, and for daily use is not likely ever to compete with the latter in the market.

It is a singular fact that this peculiar manna-sugar exists

in many familiar sea-weeds. It gives their sweet taste to those which are collected for eating along various parts of our coast, and is found in smaller quantity in many which are not perceptibly sweet to the taste. The *Laminaria saccharina*, when quite dry, contains above 12 per cent., or one-eighth part of its weight, of mannite. When the plant is dried in the air, the sugar exudes, and forms a white incrustation on its leaves. The *Halidrys siliquosa* contains from 5 to 6 per cent., and even the common *Fucus vesiculosus* 1 or 2 per cent. (STENHOUSE.) No use is made of this sugar of sea-weeds, except in so far as it assists, in some cases, in making them eatable.

Fig. 44.

Eucalyptus resinifera—The Iron Bark Gum-tree.
Scale, 1 inch to 60 feet.
Leaves, 1 inch to 5 inches.

Mannite in small quantity may also be extracted from common celery, and from the root of the dandelion; and it can be formed artificially from cane sugar.

2°. *Eucalyptus sugar*, or *gum-tree manna.*—The genus Eucalyptus, or gum tree of the colonists (fig. 44), forms a distinguishing feature in the landscape and forest scenery of Australia and Van Diemen's land. At certain seasons of the year, a sweet substance exudes from the leaves of these trees, and dries in the sun. When the wind blows, so as to shake the trees, this Australian manna is sometimes seen to fall like a shower of snow. Like the true manna, this sweet substance contains a peculiar crystallisable sugar—different, however, in composition and in some of its properties from the mannite already described. Though it is said to be produced in considerable quantities, I have not learned that it is customary to collect it for use as a sweet, either in Van Diemen's land or in Australia.*

3°. *Other mannas.*—Other sweet substances also are obtained from plants, to which the name of manna has been given. Thus, oak manna exudes from the leaves of a species of oak common in Kurdistan, and known to botanists as the *Quercus mannifera*, or manna-bearing oak. Larch manna is a sweet substance, which, in some countries, is found upon the European larch (*Larix Europœa*) about the month of June. Cedar manna occurs in small globules on the branches of the *Pinus cedrus*. It is brought from Mount Lebanon, where it sells as high as 20s. or 30s. an ounce. It is much esteemed in Syria as a remedy for affections of the chest. Persian manna, or *Gen*, called also Alhagi manna, and by the Arabs *Tereng jabim*, is obtained from the camel's thorn (*Hedysarum alhagi*, Linn.), a plant which is indigenous over a large portion of the East. It yields manna, however, only in Persia, Bokhara, Arabia, and Palestine. Extensive plains are in these countries covered with the alhagi, and it is of great importance as food for the camels, as well as for sheep and goats. From the wounds produced by the browsing of these animals the manna chiefly exudes. It is collected by the Arabs and caravans which cross the Desert, and is used as food. It is gathered by merely shaking the branches.

Tamarisk manna is obtained from the *Tamarix mannifera*, a tree which grows abundantly in the neighbourhood of Mount Sinai. The manna of the Old Testament is sup-

* See the Author's *Lectures on Agricultural Chemistry and Geology*, 2d edition, p. 181.

posed by some to have been that of the camel's thorn, and by others that of the tamarisk. Both trees grow in the wilderness of Sin, along certain parts of the route of the ancient Israelites, and both yield limited supplies of a sweet manna. If the produce of either of these trees was the true manna of the Israelites, the miracle by which they were so long fed with it consisted—*first*, in a wonderful multiplication of the produce, so as to sustain millions where probably not a score of persons could be sustained on the quantity naturally produced; and, *second*, in causing it to follow and fall daily around them in parts of the wilderness where none of the trees grow, and in equal abundance all the year around. That is to say, the sustenance of the wandering people was the result of a constant miracle, whether the manna was of a kind which might or might not have been derived from either of these natural sources.

Fig. 45.

Tamarix gallica mannifera—The Manna-bearing Tamarisk.
Scale, 1 inch to 12 feet.
Flowering branch, 1 to 5 inches.

In the Wady Feiran—the valley which leads from the Gulf of Suez towards Mount Sinai—the traveller passes through thick avenues of Turfeh or Tarfa trees (*Tamarix mannifera*, fig. 45), bending over his head like the alleys of a garden. This tree resembles the weeping birch, but is still more delicate in appearance, and the so-called manna flows in drops from the extremities of its slender pensile boughs. A small quantity is collected and carried to the convent of Sinai, where it is prepared by boiling and put into small tin cases, which are disposed of to pilgrims and other visitors. In this state it resembles melted gum with small rounded grains in it, and has a somewhat similar taste, only sweeter and rather aromatic." * The manna is supposed to flow in consequence of the puncture of the *Coccus maniparus*, an insect which infests the tamarisk trees. It exudes as a thick syrup, which, during the heat of the day, falls in drops, but during the night congeals, and is gathered in the cool of the morning. Its solution in water readily ferments. It is eaten in Palestine and about Sinai as a delicacy, and, like the cedar manna, is esteemed as a remedy in diseases of the chest. The total quantity of this manna now collected in the desert of Sinai appears to be comparatively trifling.

Dr. Milman and Dr. Lepsius both regard this sweet substance as the manna of Scripture, and consider its properties to be generally the same as those ascribed by Moses to that collected by the children of Israel. Dr. Robinson, on the other hand, denies that their properties at all correspond. I agree with Dr. Robinson. In doing so, however, I do not lay so much stress on alleged differences in taste, in general appearance, &c., as on the very remarkable property mentioned in the following passage :—

"And Moses said, Let no man save of it till the morning. Notwithstanding they hearkened not unto Moses, but some of them left of it till the morning, and it bred worms and stank, and Moses was wroth with them."—(Exodus, xvi. 19, 20.)

This rapid putrefaction, the smell, and the breeding of worms, are properties which belong to no known variety of sweet vegetable exudation. It implies something of an ani-

* Bartlett's *Forty Days in the Desert*, p. 68. The figure I have given does not represent the graceful tree described by Bartlett. It varies in appearance in different localities, and I cannot find that any representation of the entire tree has anywhere been published. In a book so beautiful as Mr. Bartlett's one might have expected to find this tree, which he describes so graphically.

mal nature, or the presence in considerable quantity of a substance analogous to the gluten of plants or the fibrin of animals.* And the presence of such a substance, again, accounts for the very nutritious qualities ascribed to this manna, and which is so superior to that of any other vegetable sweet with which we are acquainted. The manna of Scripture, therefore, I believe to be still unknown, as well as the immediate or natural source from which it might have been derived.

Orcin manna.—Orcin is a sweet substance which exists in certain species of lichen. By Berzelius it was named Orcin sugar, because of its sweetness; and by Robiquet it was regarded as a variety of manna. In chemical composition and properties, however, it is very different from any of our common sweets, and it has a disagreeable after-taste, which would alone prevent it from finding a place among the luxuries of life.

IV. Liquorice Sugar.—The root of the common liquorice (*Glycyrrhiza glabra*), fig. 46, contains a peculiar sweet substance, which, when extracted with water, has the property of becoming dark-coloured or black in the air. The dried extract is known in this country under the names of Spanish and Italian juice, from the countries in which it is most abundantly produced. It differs in flavour from all the other sugars I have mentioned; it does not crystallise, and it does not ferment when yeast is added to it.

For medicinal purposes the root is largely cultivated at Mitcham in Surrey, and other places. The extract is imported partly in the sticks, known under the name of Spanish Liquorice; and partly in solid masses, run into boxes containing about two hundredweight each. In 1850, about 500 tons were imported. It does not compete directly,

* See The Beef we cook.

however, with cane sugar. A considerable quantity, no doubt, is eaten as a sweet, and to give relief to affections of the throat, but the principal consumption is said to be by the brewers in the manufacture of porter.

Fig. 46.

Glycyrrhiza glabra—The Liquorice plant.
Scale, half an inch to a foot.

The roots of *Glycyrrhiza echinata*, *G. glandulifera*, of *Trifolium alpinum*, and of *Abrus precatorius*, are said to possess the same properties as the common liquorice; and among other sweets which resemble that of liquorice, is one which is found in the root of the *Ononis spinosa*. To this variety the discoverer has given the name of Ononid. It is not likely, however, to become of any economical importance.

V. Milk Sugar.—Milk contains a peculiar species of sugar, to which the sweetness of milk is owing. When the curd is separated in the making of cheese, the sugar remains in the whey, and may be obtained in the form of crystals by boiling the whey to a small bulk, and setting it aside to cool. This sugar is hard and gritty when crushed between the teeth, is less soluble and less sweet than cane sugar. In Switzerland, and some other cheese countries, it is extracted for sale, but the manufacture and consumption of milk sugar is on the whole very trifling. In plants it rarely occurs—the acorn being almost the only common vegetable production in which it has, as yet, been detected.

Among the most important of the varieties of sugar above described—the grape, fruit, cane, and milk sugars—there exists a remarkable analogy in chemical composition. They all consist of the three elementary bodies already described under the names of Carbon, Hydrogen, and Oxygen.* And in all of them the hydrogen and oxygen are in the proportions to form water, so that we can, for simplicity of language, say, that they are composed of carbon and water. The proportion of this water is not the same in each variety of sugar, neither is it always different. Thus—

36 lb. of carbon, and	54	of water, form	90	of crystallised cane sugar.
36	„	63	„	99 of grape or fruit sugar.
36	„	54	„	90 of milk sugar.

Thus, in the larger proportions of water it contains, we seem to see a reason for the difference in sweetness, and other properties which grape sugar exhibits when compared with cane sugar. But on the other hand, the proportions of carbon and water in crystallised cane and milk sugars are identical, and yet between these two kinds of sugar, the difference of properties is equally great. This last is a very remarkable circumstance, and presents the first example, which has fallen in our way, of one of the most interesting discoveries of modern chemistry—that two compound substances may consist of the same elementary bodies united together in the same proportions, and yet be very different from each other in their properties.

Other kindred illustrations of this principle are presented by the woody or cellular fibre (cellulose), the starch, and the gum, which, as I have explained (p. 201), may be artificially converted into grape sugar by the action of weak sulphuric acid. Thus—

36 lb. of carbón united to 45 lb. of water, form 81 lb. either of cellulose, of starch, or of gum.

* See chapters I. and II.—THE AIR WE BREATHE, AND THE WATER WE DRINK.

And yet each of these three substances is very different in its properties from either of the other two.

Again, the dark-brown vegetable matter (humic acid) to which the colour of soils is partly owing, consists of carbon and water only, for

36 of carbon, and 27 of water, form 63 of humic acid.

Now, in regard to substances so composed, it is not difficult, with the aid of this knowledge, to form a general idea of the way in which they may be transformed, one into the other. Thus—

63 of humic acid united to 18 of water, *may* form 81 of cellulose, starch, gum, or sugar.
81 of starch, with 9 of water, *may* form 90 of cane sugar.
90 of cane sugar, with 9 of water, *may* form 99 of grape sugar.

And changes of this kind really take place in nature. Thus the humic acid of the soil enters the roots of plants, and in the interior of the plant is changed into the cellulose or woody matter of its growing shoots, and into the starch of its seeds. The starch of the tasteless pear, of the banana, and of the bread fruit (p. 96), changes into sugar as the fruit ripens and becomes sweet. And by the action of acids in the sour saps of plants, and in somewhat acid fruits, cane sugar, which is first produced, is changed into grape sugar. In all these cases, the substance which disappears only combines with a little more water, to form the new compound which is produced.

And we artificially imitate these natural operations when, in the manufacture of potato sugar, we transform the starch of the potato into a sweet resembling the sugar of grapes, or when, by the prolonged action of sulphuric acid, we change sawdust or rags into a similar sweet.

In these changes, the acid employed possesses the singular property of causing the carbon of the starch or woody fibre to unite with a larger proportion of the elements of water, and thus to assume the form of grape sugar. And it is out of such observed *reactions* of bodies—as such influences are called—that new chemical arts are daily springing up. Thus the manufacture of potato sugar, already described, is a valuable independent art, founded solely upon a knowledge of this action of sulphuric acid. But many other arts, besides, have been either wholly based upon, or have been greatly improved, by the application of this property I instance only the manufacture of a dye-stuff called *garancine.*

Madder, as is well known, is the root of a plant (*Rubia tinctorum*) which is cultivated largely in certain parts of Europe, and the Levant, for the sake of the beautiful red colours it gives to the fibres of cotton and wool. This root, when dried and ground to fine powder, is the common madder of the dyer. But, besides the valuable colouring matter, this root contains gum, gluten, mucilage resembling that obtained from Iceland moss, and various other substances, which interfere with its use as a dye, and render the use of it difficult to the dyer, and the colour it imparts in some degree uncertain. In the course of the many chemical investigations to which this substance has been subjected, however, it was observed, that while sulphuric acid, under certain circumstances, acted upon nearly all these useless parts of the root, it had no effect upon the colouring matter. The former it changed into easily soluble sugar, or altogether destroyed; while to the latter it only gave new brightness and beauty. The application of this was obvious. The ground root was steeped for so many hours in sulphuric acid mingled with so much water, and was then washed perfectly free from acid, and again dried. It was now the colouring matter, or *garancine*, comparatively pure—in some cases 5, but usually about 3 times more powerful as a dye than the natural root. It was less bulky and lighter for carriage in proportion, was more easy to use, and more certain in the shades of colour it gave to cloth.

Thus, from the application to madder root of the observed action of sulphuric acid upon vegetable substances allied to our sugars, arose both the new art of making *garancine*, and important improvements in the old art of dyeing.

Thousands of similar reactions are known to chemists; and the origin of almost every art of life may be traced to the first observation of some one of the countless visible influences which one form of matter exercises over another.

Melted soda dissolves sea-sand, and the solution, when cold, is our common window-glass. Hence the magnificent glass-trade of our time.

Potash melted with hoofs and horns, and thrown carelessly into water containing iron, gave an intense blue colour. This was Prussian blue; and hence a crowd of arts and manufactures, and of beautiful applications of chemistry, have sprung up.

Every day new arts sprout up, as it were, beneath our feet, as we linger in our laboratories observing the new reactions of probably new bodies; and in each new art is seen a new means of adding to the comforts and luxuries of mankind, of giving new materials and facilities to commerce, and of increasing the power and resources of nations.

For pleasing examples of such arts—just bursting into leaf like the buds before our eyes in the sunshine of our English spring—I refer the reader to a succeeding chapter on THE ODOURS WE ENJOY.

CHAPTER XII.

THE LIQUORS WE FERMENT.

THE BEERS.

Our fermented drinks.—Grape sugar is changed into alcohol by fermentation.—Cane sugar and starch converted into alcohol.—Production of diastase during the sprouting of corn.—Action of this substance upon starch.—How the infant plant is fed.—Malt liquors; principles involved in the preparation of.—The malting of barley.—The making of beer.—Influence of diastase on the processes.—The fermentation of the wort.—Influence of the yeast.—How the yeast plant grows and multiplies; its remarkable influence still inexplicable.—Composition of beer.—Proportions of malt extract and of alcohol.—Beer characterized by its nutritive quality and its bitter principle.—Chica or maize beer of South America.—Maize malt—Preparation of chica mascada or chewed chica.—How the chewing promotes the process and gives strength to the chica.—Influence of the saliva.—Chica from other vegetable substances.—Bouza or millet beer of Tartary, Arabia, and Abyssinia.—Murwa beer of the Himalayas.—Chemical peculiarities of these millet beers.—Quass or rye beer.—Koumiss or milk beer; mode of preparing it; its composition and nutritious qualities.—Lactic acid in this beer.—Ava, cava, or arva.—Extensive use of this drink among the South Sea Islanders; how it is prepared and used; its narcotic qualities.—Effect of chewing on the ava root.—Ceremonies attending its preparation and use in the Tonga and Feejee islands.

THE liquors we ferment are all directly produced, either from the natural sugars which we extract from plants, or from the sugars which we prepare by art. I shall briefly advert to the most interesting and important of these liquors now in use in different parts of the world. The way in which these drinks are prepared, their chemical composition, and their chemico-physiological action upon the system, are more or less connected with the common life of almost every people.

I. The Beers.—When grape sugar is dissolved in water, and a little yeast is added to the solution, it begins speedily to ferment. During this fermentation, the sugar is split up into three different substances—alcohol, water, and carbonic acid.* The two former remains in the liquid while the carbonic acid gas escapes in bubbles into the air.

When common cane sugar is dissolved in water and mixed with yeast in a similar way, fermentation is induced as before. The cane sugar is first changed into grape sugar by the action of the yeast, and then the grape sugar is split up into alcohol, water, and carbonic acid. These changes take place in close as well as in open vessels, so that the presence of air is no way necessary to their perfect and rapid completion.

If starch be converted into grape sugar by the action of diluted sulphuric acid, or of a mixture of malt, as described in a preceding chapter,* and yeast be then added to the sweet solution, the same changes and the same production

* This splitting up takes place as follows:
Let C denote carbon, H hydrogen, and O oxygen—

		C	H	O
Then *one* of grape sugar, ...	=	12	14	14
Two of alcohol, ...	=	8	12	4
Four of carbonic acid, ...	=	4	0	8
Two of water, ...	=	0	2	2
And these together make . . .		12	14	14

So that the substance of one of grape sugar is split up into two of alcohol, four of carbonic acid, and two of water. This splitting up is induced by the yeast, which, however, affords none of the materials of which the alcohol, &c., consists.

* The Sweets we extract, p. 197.

of alcohol take place. From potato starch, treated in this way, large quantities of spirit (potato brandy) are manufactured in France, Germany, and the northern countries of Europe.

But by a still more beautiful process the starch of barley and other grains is converted into grape sugar before it is removed from the seed, and is then split up as before, by means of yeast, into alcohol, water, and carbonic acid.

In a previous chapter* it has been shown that these grains consist essentially of two principal substances—starch and gluten. When moistened, in favourable circumstances, the grain begins to sprout. The starch and gluten it contains are, of course, intended to form the first food of the young plant; but these substances are insoluble in water, and therefore cannot, in their natural state, pass onwards from the body of the seed to supply the wants of the growing germ. It has been beautifully provided, therefore, that both of them should undergo chemical changes as the sprouting proceeds. This takes place at the base of the germ—exactly where and when they are wanted for food. The gluten is changed, among other products, into a white soluble substance, which has been distinguished by the name of *diastase*, and the starch into soluble grape sugar. Hence the sweetness of sprouted corn.

Starch can be transformed into sugar, as I have explained (p. 201), by the agency of a minute quantity of sulphuric acid. It is so transformed also by this diastase. Produced in the sprouting seed in contact with the starch, the diastase changes the latter into sugar, and makes it soluble in the sap just as it is required. By this means the infant plant is fed.

The maltster, brewer, and distiller, avail themselves of

* See THE BREAD WE EAT, p. 79.

this natural change in the constituents of sprouting grain, and on a large scale call into action the remarkable chemical influence of diastase. This is abundantly illustrated by the chemical history of the art of brewing.

1°. Malt Beers are so called because they are prepared, either in whole or in part, from infusions of malted barley. The manufacture of these drinks involves two distinct chemical processes: *first*, The change of the starch of the grain into sugar; and, *second*, The change of the sugar into spirit-of-wine or alcohol. With a view to the first of these ends, the grain is manufactured into malt; to attain the second, it is submitted to fermentation through the medium of yeast.

a. The malt.—The maltster moistens his barley in heaps, and spreads it on a floor in a dark room to heat and sprout. When the germ (*acrospire*, he calls it) is about to burst from the envelope of the seed, he arrests the growth by drying the grain gently on the floor of his kiln. It is now malted barley, and has a sweet taste, showing that it already contains sugar. Other grains—such as wheat, oats, and rye—may be converted into malt by a similar process. Even Indian corn is malted in North America; and in South America this malt has been used for making beer from the remotest times. In Europe, however, barley has been found by long experience to be best adapted for this process—though malted rye and wheat are employed along with the barley for the manufacture of some particular kinds of beer.

b. The beer.—The brewer or distiller bruises the malt and introduces it into his mash-tun, with water gently warmed to 157° or 160° Fahr. This water dissolves first the sugar which has already been formed in the seed, and afterwards the diastase. This latter substance then acts upon the rest of the starch of the seed, converting it first into a species of soluble gum, and finally into grape sugar. If the process has been well conducted, little but the husk of the grain is left undissolved, and the liquor or wort has a decidedly sweet taste.

Three circumstances are remarkable in regard to this diastase. *First*, That even in good malt, about one pound of diastase only is formed for every hundred parts of starch contained in the grain. *Second*, That this one pound of diastase is sufficient to change a thousand pounds of starch into grape sugar. And *third*, That by heating the solution containing it to the boiling point, the diastase is killed, as it were: its power of changing starch into sugar is wholly destroyed.

The first and second of these circumstances enable the brewer, if he choose, to mix with his malt a certain portion of starch, or of unmalted grain. The diastase of the malted portion is sufficient to transform into sugar, not only the whole starch of the malt, but all the starch also of the raw grain. Thus both the expense and the waste which would attend the malting of the latter is avoided. In this country the brewer rarely avails himself of this opportunity of adding raw grain. Continental brewers, however, and our home-distillers, both practise it largely.

The third circumstance determines the time when the wort may be safely boiled—which is the next stage in the manufacture of beer. The change of all the starch into sugar being effected, the diastase is no longer of service, and the wort may be heated to boiling, with advantage. By this higher temperature the action of the diastase is stopped, and at the same time the albumen which the water has dissolved out of the grain is coagulated and separated in flocks. Advantage is taken also of this boiling, to introduce the hops; and these, besides imparting their peculiar bitterness and aroma to the liquid, help further to clarify it. Both the length of the boiling and the quantity of hops added to the liquid vary with its richness in sugar, and with the quality of the beer it is intended to make.

The boiled liquor is run off into shallow vessels, and cooled as rapidly as possible to the best fermenting temperature, which lies between 54° and 64° Fahr. It is then transferred to the fermenting tun; a sufficient quantity of yeast is added—obtained, if possible, from the same kind of beer it is desired to make—and it is allowed to ferment slowly for six or eight days. During this fermentation, the sugar of the wort is split up into the alcohol and water, which remain in the beer, and into the carbonic acid gas which, for the most part, escapes from the surface of the liquid and mingles with the surrounding air.

Three things are notable in this process: *first*, That the quantity of yeast which is added, and the temperature at which the liquor is afterwards kept to ferment, vary with every kind of beer; *second*, That the yeast has a tendency to reproduce a beer which, in flavour, &c., shall resemble that from which it has been obtained; and *third*, That the whole of the sugar contained in the wort is never in practice transformed into alcohol. Good beer—however clear, hard, bright, and bitter—always retains a pleasant sweetish taste. From one-half to three-fourths only of the sugar in the wort is decomposed. Were the fermentation not so regulated as to leave this residue of undecomposed sugar, the beer would refuse to keep. It would turn sour in the cask.*

I do not follow further the manufacture of this important beverage. But I cannot dismiss the beautiful series of operations of which it consists, without calling the attention of my reader for a moment to the remarkable place

* URE's *Dictionary*, pp. 108, 109.

which the minute yeast plant (fig. 47) occupies among the agents by which the final result is attained. I have already described this plant; how small it is; how mysteriously it appears, and how rapidly it grows (p. 71).

Fig. 47.

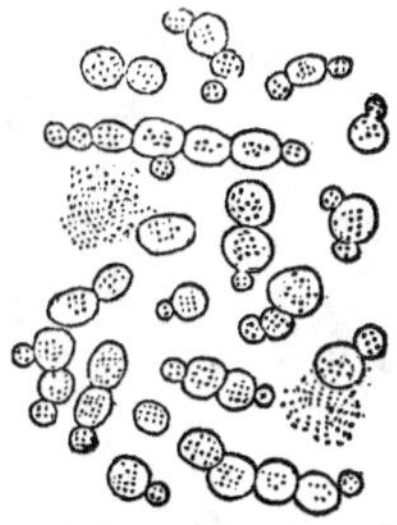

Yeast after being in wort for eight hours, showing—
The transparency of the yeast cells.
The granules or nuclei in their interior.
How the spores or seeds escape from the interior of the cells.
How they germinate and multiply by budding.
How they unite into jointed filaments.

As sulphuric acid and diastase, by mere contact apparently with starch, convert it wholly into sugar; so yeast, by a similar species of contact, converts the sugar wholly into alcohol, water, and carbonic acid. How either of these transformations is effected by the agents employed, we cannot explain.

There is this interesting difference in the way in which these three agents operate—that, while the sulphuric acid employed to transform starch into sugar remains unchanged in quantity, and while the diastase itself changes and disappears, the yeast lives, multiplies, grows, increases in quantity, and augments in size and vegetable development. The minuteness of the yeast plant, consisting in its simplest form of only a single cell, long prevented it from being generally regarded as a form of living matter. But the changes it undergoes in the fermenting tub, day by day, as shown by the microscope, prove it to be unquestionably a growing vegetable. The drawing given above (fig. 47) shows the appearance it has assumed after being in the wort only eight hours. The cells have multiplied, increased in size, and begun to string themselves together like beads. The drawing in fig. 48 exhibits a still more developed and unquestionable plant-form sometimes found in the yeast deposited by fully fermented London porter. The increase in the quantity of yeast during such fermentation is so great, that 35 lb. of dry yeast employed in brewing 1250 gallons of beer, have been known to increase to, or yield, 247 lb.

Fig. 48.

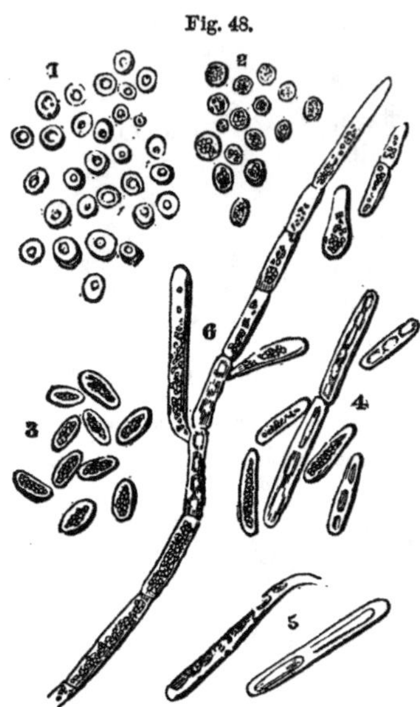

Mycoderma cervisiæ.—A developed yeast-plant.

The numbers indicate the successive stages of the growth or development.

But that the yeast lives and increases in the fermenting liquid, does not explain its action upon the sugar. The mystery remains none the less. How this plant, in growing rapidly itself, should induce the sugar at the same time to split itself up as I have described, and that without combining with or otherwise appropriating any of the new substances produced—this is still altogether inexplicable. Neither chemistry nor physiology can as yet hazard even a plausible, light-bringing conjecture upon the subject. It is something, however, to be able to see, in regard to any point that we have reached, the actual limits of our positive knowledge.

The composition of the beer, obtained as I have described, varies with almost every sample.

a. When beer is evaporated or boiled to dryness, it leaves behind a certain quantity of solid matter, usually spoken of as malt extract. This consists of undecomposed sugar, of soluble gluten from the grain, of bitter substances derived from the hop, and of a certain proportion of mineral matter. It varies in quantity from less than 4 to upwards of 8 lb. in every 100 lb. of good beer. In fine wine-like beers, such as our modern English bitter beers, the quantity of extract is small. In heavy sweet beers, it is large. Good Edinburgh ale contains about 4 per cent., or nearly half a pound to the gallon. The German Brunswick beers are remarkable in this respect. A sweet small-beer of that city contains 14 per cent. of extract; and a scarcely half-fermented black drink, called Brunswick *mumme*, as much as 39 per cent.—about 5 lb. to the gallon. The nutritive qualities of beer, which are often considerable, depend very much upon the amount and nature of this extract.

b. But beer contains alcohol also, the result of the fermentation; and this varies in quantity quite as much as the extract. Thus—

	Of Alcohol.		
Small beer contains . .	1 to 1½	per cent.	by weight.
Porter, ...	8¼ to 5¼	"	"
Brown stout, . . .	5½ to 6¼	"	"
Bitter and strong ales, . .	5¼ to 10	"	"

By measure, these proportions of alcohol are about one-fourth more than the numbers above given.

Upon this alcohol depends the purely intoxicating effect of malt liquors. And in this respect our strong ales have about the same strength and influence as hock and the light French wines. But they contain, in addition, and as distinguishing them from the wines,

First, The nutritive matters of the extract which are derived from the grain.—These, as I have said, vary from 4 to 8 per cent. In milk, the model food, the nutritive matter amounts to 12 per cent., and is, besides, somewhat richer in curd, the ingredient which corresponds to the gluten of plants. Beer, therefore, is food as well as drink. A little beef eaten with it makes up the deficiency in gluten, as compared with milk; so that beef, beer, and bread—our characteristic English diet—are most philosophically put together, at once to strengthen, to sustain, and to stimulate the bodily powers.

Second, The bitter narcotic principle of the hop.—By this, not less than by its nutritive quality, beer is distinguished from wine. Of this ingredient and its effects I shall treat in a subsequent chapter.*

2°. Chica, or Maize Beer.—The use of malt beer in Germany, and probably also in England, is very ancient; but that of chica or maize beer in South America appears to be equally remote. It was a common drink of the Indians long before the Spanish conquest.

The usual way of preparing chica is to water or moisten Indian corn, as the English maltster does his barley—to leave it till it sprouts sufficiently, and then to dry it in the sun. It is now maize malt. This malt is crushed, mashed in warm water, and then allowed to stand till fermentation takes place. The liquor is of a dark yellow colour, and has an agreeable, slightly bitter, acid taste.† It is in universal demand throughout the west coast of South America, and is consumed in vast quantities by the mountain Indians. Scarcely a single hut in the interior is without its jar of the favourite liquor.

In the valleys of the Sierra, however, the most highly-prized chica is made in a somewhat different manner. "All the members of the family, including such strangers as choose to assist in the operation, seat themselves on the floor in a circle, in the centre of which is a large calabash, surrounded by a heap of dried maize (malt). Each person takes up a handful of the grain and thoroughly chews it. This is

* See The Narcotics we indulge in.

† Von Tschudi, *Travels in Peru*, p. 151.

deposited in the calabash, and another handful is immediately subjected to the same process, the jaws of the company being kept continually busy until the whole heap of corn is reduced to a mass of pulp. This, with some minor ingredients, is mashed in hot water, and the liquid poured into jars, where it is left to ferment. In a short time it is ready for use. Occasionally, however, the jars are buried in the ground, and allowed to remain there until the liquor acquires, from age, a considerable strength, and powerfully intoxicating qualities.

Chica thus prepared is called *chica mascada*, or chewed chica, and is considered far superior to that prepared from maize crushed in the usual manner. The Serrano believes he cannot offer his guest a greater luxury than a draught of old chica mascada, the ingredients of which have been ground between his own teeth.*

Disgusting as this process of manufacture appears to the European, it is nevertheless founded in reason, and presents a sort of instinctive or experience-born application of a beautiful chemico-physiological principle.

We have seen that grain is malted in order that diastase may be produced, and that it is then bruised and digested in warm water, in order that this diastase may convert the starch into sugar. But the saliva of the mouth possesses a similar property of converting starch into sugar. Mix starch intimately with saliva, and keep the mixture moderately warm for a time, and sugar will gradually be produced.

This is what the Indian does in preparing his chica mascada. He chews the grain thoroughly: this reduces it to a fine pulp, and at the same time mixes it intimately with saliva. When set aside, this pulp sweetens and afterwards ferments.

* *The Leisure Hour*, June, 1853, p. 372.

The maize he makes his liquor from is a large grain. The diastase produced during the malting—which is not always well conducted—is often insufficient to convert the whole of the starch into sugar, but the mixture of saliva aids the diastase, and insures the change. It also aids in producing and promoting the fermentation which suceeds.

It is very interesting to discover so beautiful a chemico-physiological reason for a practice so disagreeable and apparently so unaccountable.

Chica is not always made from maize. It is prepared also from barley, rice, pease, yuccas, pine-apples, grapes, and even bread—(VON TCHUDI). The name, originally restricted to the liquor obtained from maize, appears to have been gradually applied to the fermented drinks of various kinds which are in use in different parts of South America. A variety of chica mascada is made in some places from the pods of the *Prosopis algaroba*, which are very sweet, mixed with the bitter stalks of the *Schinus molle*. Old women are employed to chew these pods and stalks. The chewed pulp is mixed with water, and the mixture soon ferments and forms an intoxicating beer.* The addition of the bitter ingredient in this case is interesting, not only because it resembles our own more recent practice of adding hops and other bitters to our beer, but because it intimates the existence of a remarkable similarity in natural taste among tribes of men most remote in situation, and most unlike in understanding and habits.

3°. BOUZA, MURWA, or MILLET BEER, is a favourite drink of the Crim Tartars. They prepare it from fermented millet-seed, to which they add certain admixtures which render it excessively astringent—(OLIPHANT †). They call it Bouza.

* *Chemical Gazette*, 1844, p. 131, note.

† *Russian Shores of the Black Sea*, p. 277.

The Arabians, Abyssinians, and many African tribes, give the same name to a fermented drink which they usually prepare from *teff*, the seeds of the *Poa Abyssinica*. They occasionally employ millet-seed, however, and even barley, for the purpose. Their bouza is described as a sour, thick drink.

In Sikkim, on the southern slopes of the lower Himalaya, millet beer, under the name of *murwa*, is in very general use. It is prepared by moistening the millet-seed (*Eleusine coracana*), and allowing it to ferment for some days. On a portion of this, considered sufficient for the occasion, or for the day's consumption, hot water is then poured. It is usually drunk while still warm—is served in bamboo jugs, and sucked through a reed. When quite fresh, it tastes "like negus of Cape sherry, rather sour." It is very weak, but in a hot day's march is described as a very grateful beverage—(HOOKER).*

With the chemical peculiarities of these different forms of millet beer we are at present unacquainted. The speciality in their preparation seems to be, that they are fermented in the grain, and not in the wort, as is the case with European beers; and that the fermentation is spontaneous, and not produced by yeast. Under these circumstances, three chemical changes will be proceeding in the moist grain at the same time :—

First, The starch of the grain will be transformed into sugar by the agency of the diastase, which is formed during the sprouting that ensues after the grain is moistened.

Second, This sugar is partly changed into alcohol by the fermentation which spontaneously commences.

Third, A part of the sugar is changed also into lactic

* *Himalayan Journals*, vol. i., pp. 285, 291.

acid, or the acid of milk, through the action of the gluten of the millet, which, during the spontaneous fermentation, possesses the peculiar property of producing this change.

The drink obtained by infusing this altered grain in water agrees with our European malt-liquors, therefore, in containing nutritive matters derived from the starch and gluten of the grain. But it differs from them in containing lactic instead of acetic acid. The Indian *murwa* differs from them also in being drunk like tea soon after it is infused, and in containing no bitter addition resembling our hop. The astringency of the bouza of the Crim Tartars seems to indicate that *they* use something in preparing it besides the fermented millet-seed.

It is a singular coincidence that the mode of infusing in hot water and sucking through a tube, practised on the Himalayas, is exactly the same as is practised in South America in preparing maté or Paraguay tea. In each of these remote districts the beverage prepared is taken hot, and is in universal use; and yet, so far as I am aware, this mode of drinking is adopted only in North-Eastern Asia and in Southern America. Is there anything more than a mere coincidence in this?

4°. Quass, or Rye Beer, a favourite Russian drink, is a sharp, acid, often muddy liquor, which, in taste and appearance, resembles some of the varieties of bouza. It is made by mixing rye-flour, and occasionally barley-flour, with water, and fermenting. It may possibly contain lactic acid, but I am not aware that its composition has yet been made the subject of special chemical inquiry.

This is one of the cases in which un-malted grain is employed in the manufacture of beer on the continent of Europe.

5°. Koumiss, or Milk Beer.—Milk, as I have explained in the preceding chapter, contains a peculiar kind of sugar, less sweet than cane sugar, to which the name of milk sugar is given. This sugar, when dissolved in water, does not ferment upon the addition of yeast; but when dissolved in the milk, along with the curd and butter, it readily ferments, is transformed into alcohol and carbonic acid, and gives to the liquor an intoxicating quality.* This fermentation will take place spontaneously, but it is hastened by the addition of yeast or of a little already fermented milk. The fermented liquid is the koumiss of the Tartars. Mare's milk is richer in sugar than that of the cow, and is usually employed for the manufacture of milk beer. It is prepared in the following manner:—

To the new milk, diluted with "a sixth of its bulk of water, a quantity of rennet, or, what is better, a sour koumiss, is added, and the whole is covered up in a warm place for twenty-four hours. It is then stirred or churned together till the curd and whey are intimately mixed, and is again left at rest for twenty-four hours. At the end of this time it is put into a tall vessel and agitated till it becomes perfectly homogeneous. It has now an agreeable sourish

* This transformation is effected, through the agency of the curd, in a way not yet clearly understood. The mere change of substance—that is, of the sugar into alcohol and carbonic acid, supposing it to be produced directly—appears very simple. Thus, C representing carbon, H hydrogen, and O oxygen:—

		C	H	O
One of milk sugar is	=	24	24	24
Four of alcohol are	=	16	24	8
Eight of carbonic acid, ...	=	8	0	16
Sum, ...		24	24	24

So that, in one of milk sugar there are exactly the materials to form four of alcohol and eight of carbonic acid. But the transfo mation is probably much more indirect and circuitous—the curd changing one portion of the sugar into lactic acid, this acid changing the rest of the milk sugar into grape sugar, and then the altered curd again in some unknown way, causing this grape sugar to ferment and split up into alcohol and carbonic acid. The non-chemical reader will understand in some degree, from this example, how difficult it is to follow, and distinctly make out, the rapid and successive changes which often take place in consequence of the mutual re-actions of different chemical substances.

taste, and, in a cool place, may be preserved for several months in close vessels. It is always shaken up before it is drunk. This liquor, from the cheese and butter it contains, is a nourishing as well as an exhilarating drink, and is not followed by the usual bad effects of intoxicating liquors. It is even recommended as a wholesome article of diet in cases of dyspepsia or of general debility."

By distillation, ardent spirits are obtained from this koumiss, and, when carefully made, a pint of the liquor will yield half an ounce of spirit. To this milk-brandy, when only once distilled, the Kalmucks give the name of *arraca*, and from the residue in the still they make a kind of hasty-pudding.

The Arabians and Turks prepare a fermented liquor, or milk beer, similar to the koumiss, which the former call *leban* and the latter *yaourt*. In the Orkney Islands, and in some parts of Ireland and of the north of Scotland, butter-milk is sometimes kept till it undergoes the vinous fermentation and acquires intoxicating qualities.

This milk beer has never, I believe, been chemically investigated; but we know, *first*, That it agrees with the malt beers in containing a considerable proportion of nutritive matter. The butter and cheese of the milk remain as nutritious ingredients of the beer. *Second*, That it differs from the malt beers in containing more acid, and in owing its sourness not to acetic acid but to the peculiar acid of milk, the lactic acid. In both these respects it agrees remarkably with millet beer. We shall see in the next chapter that, in the kind of acid it contains, milk beer agrees also with cider.

6°. Ava, Cava, or Arva.—Similar to chica in the mode of preparation is the ava or cava of the South Sea Islands. This liquor is in use over a very wide area of the Pacific

Ocean, and among the inhabitants of very remote islands. In Tahiti, the use of it is said to have swept off many of the inhabitants. In the Sandwich Islands it was some years ago forbidden—(SIMPSON). In the Samoan group it is the only intoxicating liquor known, and old and young, male and female, are very fond of it—(WILKES). In the Tonga Islands it is prepared and drunk on every festive occasion —(MARINER). And in the Feejee Islands, the preparation of the morning drink of this liquor for the king is one of the most solemn and important duties of his courtly attendants —(WILKES).

The name of ava is given to the root of the intoxicating long-pepper (*Macropiper methysticum*), fig. 49, which is

Fig. 49.

Macropiper methysticum.—The Ava Pepper shrub.
Scale, 1 inch to 3 feet.
Leaf, 1 inch to 2 inches. Outline of leaf, natural size.
Part of stem and root, showing section, natural size.

chewed, either in the fresh or in the dried state, as the Indian chews his maize.* The pulp is then mixed with cold water, which after a brief interval is strained from the chewed fibre, and is ready for use. The taste, to one unaccustomed to it, is not pleasant. It reminded Captain Wilkes of the taste of rhubarb and magnesia! According to the white persons who have tried it, this infusion does not intoxicate in the same manner as ardent spirits. It more resembles opium in some of its effects, producing a kind of temporary paralysis, tremors, indistinctness, and distortion of vision, and a confused feeling about the head.

The presence of a narcotic ingredient in the root of this plant is very probable. Its leaf is used very largely for chewing with the well-known betel-nut,† and is believed to have a share in producing the pleasing state of mild excitement in which the betel-chewer delights. The extraction of this narcotic substance, during the process of mastication and straining, accounts for the intoxicating qualities acquired by the liquor, before ordinary fermentation and the production of common alcohol has had time to begin. Still, that the saliva produces a chemical change in the ingredients of the root, upon which change their intoxicating quality in some measure depends, is in itself very probable, from what we know of the general properties of saliva. And the probability of such a change becomes greater, when it is considered that the intoxicating qualities of the leaf only become sensible to the betel-chewer as the roll he chews becomes softened in his mouth, and saturated with saliva.

In the Tonga Islands, the ava root, when dry, is split up into small pieces with an axe or other sharp instrument,

* Fig. 49 represents the leaf and a section of the root of the ava pepper. I have been unable to procure a figure of the entire fresh root and plant.

† See THE NARCOTICS WE INDULGE IN.

is scraped clean, and is then handed to the attendants to be chewed. No one offers to chew it but young persons who have good teeth, clean mouths, and have no colds. The women often assist—(MARINER). But as the most curious passage I have met with in connection with the preparation and use of this liquor, I quote the following from Captain Wilkes:—

"The ceremony attending the ava-drinking of the king at Somu-somu, one of the Feejee islands, is peculiar. Early in the morning, the first thing heard is the king's herald, or orator, crying out in front of his house, 'Yango-na ei ava,' somewhat like the muezzin in Turkey, though not from the house-top. To this the people answer, from all parts of the koro,* 'Mama' (prepare ava). The principal men and chiefs immediately assemble together from all quarters, bringing their ava bowl and ava root to the mbure, where they seat themselves to talanoa, or converse on the affairs of the day, while the younger proceed to prepare the ava. Those who prepare the ava are required to have clean and undecayed teeth, and are not allowed to swallow any of the juice, on pain of punishment. As soon as the ava root is chewed, it is thrown into the ava bowl, where water is poured upon it with great formality. The king's herald, with a peculiar drawling whine, then cries, 'Sevu-rui-a-na' (make the offering). After this a considerable time is spent in straining the ava through cocoa-nut husks; and when this is done, the herald repeats with still more ceremony, his command, 'Sevu-rui-a-na.' When he has chanted it several times, the other chiefs join him, and they all sing, 'Mana endina sendina le.' A person is then commanded to get up and take the king his ava, after which the singing again goes on. The orator then invokes their principal god, Tava-Sava, and they repeat the names of their departed friends, asking them to watch over and be gracious to them. They then pray for rain, for the life of the king, the arrival of wangara papalangi (foreign ships), that they may have riches, and live to enjoy them. This prayer is followed by a most earnest response, 'Mana endina' (amen, amen). They then repeat several times 'Mana endina sendina le.' Every time this is repeated, they raise their voices until they reach the highest pitch, and conclude with 'O-ya-ye,' which they utter in a tone resembling a horrid scream. This screech goes the rounds, being repeated by all the people of the koro, until it reaches its farthest limits, and, when it ceases, the king drinks his ava. All the chiefs clap their hands with great regularity while he is drinking; and after he has finished his ava, the chiefs drink theirs without any more ceremony. The business of the day is then begun. The people never do anything in the morning before the king has drunk his ava. Even a foreigner will not venture to work or make a noise before that ceremony is over, or during the preparation of it, if he wishes to be on good terms with the king and people." *

It will strike the reader as a singular circumstance, that this mode of preparing fermenting drinks—the ava and the chica—by chewing the raw materials, should exist in the islands of the Pacific, and amid the sierras of South America, and there only. The materials employed in the two regions are very different, and the chemical changes produced by the chewing in the two cases very different also, though the apparent result, in the production of an intoxicating liquor, is the same. Where did the custom originate? Is its origin continental or insular? Is it in any way connected with the eastward migrations,

* WILKES' *United States' Exploring Expedition*, vol. ii., p. 97.

which the unknown past has doubtless witnessed, towards the Pacific shores of the American continent? Where analogies of tongue and features fail, may not the occurrence of strange customs point to old national relations which now no longer subsist?

CHAPTER XIII.

THE LIQUORS WE FERMENT.

THE WINES.

The wines.—Apple and pear wines.—Cider and perry.—Differences in quality.—Varieties of cider apple.—Composition of cider; tendency to sour.—Grape wines.—Rapid fermentation of grape juice.—Circumstances influence the quality of wine.—Composition of wine.—Proportion of alcohol in different wines; proportion of sugar.—Tartaric acid the peculiar acid of grape wine.—Proportions of acid in different wines.—Œnanthic ether gives the vinous flavour to wines.—Peculiar odoriferous principles which impart to each wine its own flavour or bouquet.—Consumption of wine in the United Kingdom.—Palm wine or toddy.—How extracted from the cocoa-nut tree, and from the date tree.—Extensive use of palm wine.—Sugar-cane wine, or guarapo.—Pulque, or agave wine.

II. THE WINES.—Wines are distinguished from beers chiefly by three characters: *First*, They contain little of that solid nutritious matter which enables our home-brewed beer to feed the body as well as quench the thirst and exhilarate the spirits. *Second*, They are free from any bitter or narcotic ingredient, such as the hops we add so largely to many of our English ales. *Third*, They are all fermented, without the addition of yeast, by a spontaneous fermentation; and in consequence they contain other acids besides the acetic acid, or vinegar, to which sour beer owes its acidity.

1°. APPLE AND PEAR WINES.—Cider and perry are

well-known fermented drinks. The former especially is largely prepared and consumed in England, France, and North America.

The expressed juices of the apple and the pear contain grape sugar already formed. When left to themselves they soon begin to ferment, without the addition of yeast; and during this fermentation, the sugar is converted into alcohol in the way already described.

Cider differs in flavour, in acidity, in strength, and consequently in quality, with many circumstances. The kinds of apples which are grown and used for the purpose, the degree of ripeness they are allowed to attain before they are gathered, the time given them to mellow or ferment before they are crushed, the skill with which the several varieties are mixed before they are put into the mill, the nature of the climate, the character of the season, the quality of the soil, the mode in which the trees are managed—all these circumstances materially affect the quality of the expressed juice as it flows from the crushing-mill; and then the after-treatment of the juice may introduce a hundred new shades of difference among the several ripe ciders produced from the same juice.

In Normandy, not less than five thousand differently-named varieties of the acid or bitter apple are known, and grown for the manufacture of cider! Some of these varieties are distinguished by as many as eighteen different names in different parts of the country. In that province also it is remarked, that the cider produced upon chalk soils, from the same varieties of apple, differs in flavour from that of sandy districts, and both from that of clay soils; so that the flavour of the soil (*gout de terrain*) is in Normandy a familiar expression in reference to the qualities of this fermented drink.*

* See the author's *Notes on North America*, vol. 1. p. 170.

Amid these differences in quality, however, there are certain general chemical characters in which all ciders agree. They contain little extractive or solid nutritious matter. No bitter or narcotic ingredient has been added to them. They contain, on an average, about nine per cent. of alcohol —thus resembling in strength the common hock, the weaker champagnes, and our stronger English ales. They are also chemically distinguished from malt liquors by containing lactic instead of acetic acid. In this latter respect they agree with the spontaneously-fermented bouza, or murwa beer of Abyssinia and the Himalayas, and with the milk beer of the Tartarian steppes.

Cider is further distinguished by the great facility with which it becomes sour, or runs to acid. Hence the frequency of hard cider, the difficulty of transporting it unchanged from place to place, and the frequent disappointments which attend the efforts to keep it sound for any length of time.

2°. Grape Wine.—The name of wine is usually given among us, by way of eminence, to the fermented juice of the grape. This juice, like that of the apple, contains grape sugar ready formed; and, like the juices of the apple, the pear, the gooseberry, and most other fruits, it enters easily and speedily into spontaneous fermentation. Within half an hour, in ordinary summer weather, the clearest juice of the grape begins to appear cloudy, to thicken, and to give off bubbles of gas. Fermentation has already commenced; and within three hours a distinct yellow layer of yeast has collected on the surface, and a sensible quantity of alcohol has been formed in the body of the liquid. It is still a mystery in what way the germ, seed, or sporule of the yeast plant obtains admission into the liquid juice, and in such quantity as to give rise to an almost instantaneous fermentation.

Grape wine differs in composition and quality with a thousand circumstances. The climate of the country, the nature of the season, the soil of the locality, the variety of grape, the mode of culture, the time of gathering, the way in which the fruit when gathered is treated and expressed, the mode of fermenting the juice or *must*, the attention bestowed upon the young wine, the manner in which it is treated and preserved, the temperature at which it is kept, the length of time it is preserved,—upon these, and numerous other conditions, the composition and quality of wine are dependent. All grape wines, however, contain—

a. A notable proportion of alcohol, or pure spirit of wine. This proportion is different in different kinds of wine, and varies considerably also in wines of the same kind. Thus the proportion of absolute alcohol, by measure, in our best-known wines is as follows :—

	In 100 measures.		In 100 measures.
Port, ...	21 to 23	Rhenish, ...	8 to 13
Sherry, ...	15 „ 25	Moselle, . . .	8 „ 9
Madeira, ...	18 „ 22	Malmsey, . . .	16
Marsala, ...	14 „ 21	Tokay, ...	9
Claret, ...	9 „ 15	Champagne, . . .	5 „ 15
Burgundy, . . .	7 „ 13		

The wines we commonly drink in this country are, therefore, two or three times stronger in spirit than those of France or Germany.

b. A more or less sensible quantity of grape sugar, which has escaped the decomposing action of the fermentation. This gives to wines their sweet taste and *fruity* character. Wines are called *dry* when they contain little sugar. The order of sweetness in certain wines, as they are brought to the English market, is as follows, (JONES) :—

Claret, Burgundy, Rhine, and Mosello wines contain no sensible quantity of sugar.

Sherry	contains	4 to 20	grains	in the ounce.
Madeira	„	6 „ 20	„	„
Champagne	„	6 „ 28	„	„
Port	„	16 „ 34	„	„
Malmsey	„	56 „ 66	„	„
Tokay	„	74	„	„
Samos	„	88	„	„
Paxarette	„	94	„	„

The four last-named are called sweet wines, and the extreme fruitiness of some port wines is indicated by the large proportion of sugar which this variety of wine sometimes contains. Sugar is added to the juice of the champagne grape by the grower. This is necessary, not only to give it body, but to keep it sparkling, and to prevent its becoming sour. And it is remarkable that the selection of the kind of sugar which is added has great influence upon the flavour of the wine. If doubly-refined cane and beet sugars be added respectively to the same champagne, the one will give the liquor the aroma and pleasant flavour of the cane-juice, the other the disagreeable *gout* of the beet-root. In the wine, the senses of taste and smell readily discover traces of impurity derived from the sugar, which neither eye, nose, nor mouth can detect in the purified sugar itself.

c. A variable proportion of free acid, which imparts to them a more or less distinctly sour taste. We have seen that neither malt, beer, nor cider are ever quite free from acid, and the same is the case with wine. Only the grape wine is made sour by tartaric acid.* Thus—

Acetic acid (vinegar) is the acid of *malt beer.*
Lactic acid is the acid of *millet beer, milk beer, and cider.*
Tartaric acid is the acid of *grape wine.*

* *Tartaric acid* is the acid which gives its sourness to cream-of-tartar, and which we use along with soda in making artificial seidlitz powders. It is so named, because it is extracted from the tartar or crust which deposits itself on the sides of wine-casks or bottles, by long standing.

In all the three liquors, acetic acid is present in greater or less quantity, as this is always produced when the fermentation of alcoholic liquors is allowed to proceed too far. But lactic acid is found neither in malt beer, nor in grape wine, in sensible quantity; nor is tartaric acid found in beer or cider. These acids, therefore, charácterise the liquors in which they especially exist, and establish a marked chemical distinction among the three classes of fermented drinks to which they severally belong.

Wines made from unripe grapes sometimes contain another peculiar acid which resembles the acid of lemons (citric acid), but this acid disappears from the fruit as it ripens.

Tartaric acid exists in the juice of the grape in combination with potash, forming what is called bi-tartrate of potash, or cream-of-tartar—a substance which has a well-known sour taste. When the fermented juice is left at rest, this bi-tartrate gradually separates from the liquor, and deposits itself as a crust or tartar on the sides of the casks and bottles. Hence by long keeping good wines become less acid, and every year added to their age increases, in proportion, their marketable value.

In regard to acidity, our common wines arrange themselves in the following order:—

Sherry is the *least acid.*
Port comes next.
Champagne "
Claret "
Madeira "
Burgundy "
Rhine wines "
Moselle is *most acid.*

d. A minute proportion of an ethereal substance to which the name of *œnanthic* ether is given, and to which grape wines owe the agreeable vinous odour which characterises them all. When obtained in a separate state this ether is a very fluid liquid, of a sharp, disagreeable taste, but having an odour of wine so excessively powerful as to be almost intoxicating. It does not exist in the juice of the grape, but is produced during the fermentation. It seems also to increase in quantity by keeping, as the odour of old wines is stronger than that of new wines. So powerful is the odour of this substance, however, that few wines contain more than one-four-thousandth part of their bulk of it! Yet it is always present, can always be recognised by its smell, and is one of the general characteristics of all grape wines.

e. Besides the general vinous flavour derived from this œnanthic ether, all wines contain one or more odoriferous, more or less fragrant, substances, to which the peculiar *bouquet* or scent of each is due. As these give the special character to the wine, they are more or less different in each variety. They are present even in more minute quantity than the œnanthic ether, and their chemical nature is as yet very little understood.

Grape wine is the principal fermented drink of the southern European nations. The consumption in the United Kingdom in 1853 amounted to upwards of seven millions of gallons (7,197,572). This is chiefly consumed by the upper classes. In England, beer is the poor man's substitute, while in Scotland and Ireland, whisky, more or less diluted with water, takes its place.

3°. Palm Wine, or Toddy.—The sap of many palm-trees is rich in sugar. In some countries this sugar is extracted by boiling down the collected juice, as cane sugar is extracted from the expressed juice of the sugar cane (*see* p. 219). In other countries the juice is allowed to ferment, which it does spontaneously, and in hot climates within a very short period of time. This fermentation converts the sugar into alcohol, and the juice which contains it into an intoxicating liquor.

In the islands of the Indian Archipelago, the Moluccas, and the Philippines, an intoxicating liquor is prepared in this way from the sap of the gommuti palm, *Saguerus saccharifer*. It is called *neva* in Sumatra, and the Batavian arrack is distilled from it. The cocoa palm, *Cocos nucifera* (fig. 50), produces the palm wine, known in India and the Pacific by the name of *toddy*. The mode of collecting it in the islands of the Pacific is thus described by Capt. Wilkes:—

Fig. 50.

Cocos nucifera—The Cocoa-nut Palm.
Scale, 1 inch to 12 feet.

"The karaca or toddy is procured from the spathe of the cocoa-nut tree, which is usually about four feet long and two inches in diameter. From this spathe the flower and fruit are produced; but in order to procure their favourite toddy, it is necessary to prevent nature from taking her course in bringing forth the fruit. With this view they bind up the spathe tightly with sennit, then cut off the end of the spathe and hang a cocoa-nut shell to catch the sap as it exudes. One tree will yield from two to six pints of karaca. When first obtained from the tree it is like the milk of the young cocoa-nut, and quite limpid, but after it stands for a few hours it ferments and becomes acid. When the sap ceases to drop, another piece is cut off the spathe, and every time the flow ceases the same process is repeated until the spathe is entirely gone. Another spathe is formed soon after, above this, which is suffered to grow, and when large enough is treated in the same manner." *

Fig. 51.

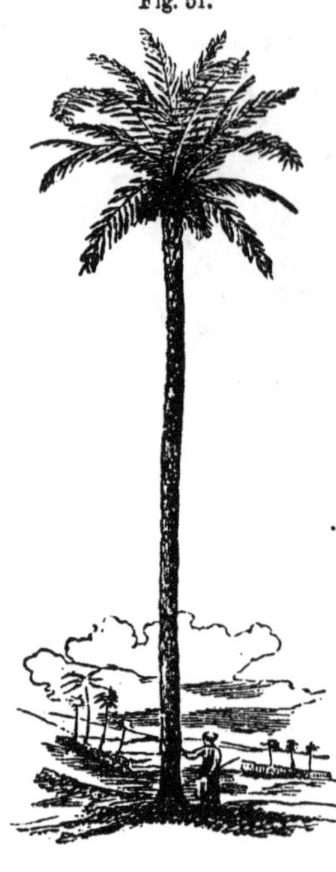

Phœnix dactylifera—The Date Palm.
Scale, 1 inch to 20 feet.
Fruit, 1 inch to 2 inches.

This method of cutting the spathe, or flowering head, is a very common one for procuring the sweet sap of the palm trees. In some countries, however, it is obtained, like that of the sugar maple and the manna ash, by simply making an incision near the top of the tree. This custom prevails in the interior of Africa, and in the Indian province of Bahar, where the abundant date-palm (fig. 51) is yearly bled for the favourite toddy. Dr. Hooker thus describes a grove

* *United States' Exploring Expedition*, vol. ii., p. 220.

of date palms in which he encamped on the banks of the Soane river in that province:—

"All were curiously distorted, the trunks growing zig-zag, from the practice of yearly tapping the alternate sides for toddy. The incision is made just below the crown, and slopes upwards and inwards. A vessel is hung below the wound, and the juice conducted into it by a little piece of bamboo. This operation spoils the fruit, which, though eaten, is smaller and much inferior to the African date." *

In India, generally, it is the fan palm (*Borassus*) which is chiefly bled for toddy. But in Bahar the date tree is preferred, because its sap more readily ferments. In the fertile oases which are sprinkled over the desert Sahara of Northern Africa, where date-tree forests cover the soil, and form the chief food and wealth of the inhabitants, this variety of palm is constantly tapped in the flowering season by the Arab and other Mahommedan tribes. They call the sap *lagmi*, and from two to three pints are yielded by each tree in a single night. But wine of the best quality is said to be yielded by the oil palms (*Cocos butyracea* and *Elais guineensis*) which grow on the west African coast; while for abundant yield few excel the *Caryota urens*, the most beautiful of Indian palms, which will often yield a hundred pints of toddy in the twenty-four hours!—(ROXBURGH.)

In the oasis of Tozar, a dependency of Tunis, the date wine is to be found in every house, and reeling Arabs are frequently to be seen in the streets of its principal towns. They are strict Mahommedans; but they justify their apparent disobedience to the Prophet by saying, "Lagmi is not wine, and the Prophet's prohibition refers to wine." †

The juice of the palm tree varies in quality with the spe-

* *Himalayan Journals*, vol. i., p. 35.

† *Evenings in my Tent.* By the Rev. WILLIAM DAVIS.

cies of palm, and with the locality in which it is grown. No chemical examination of it has yet, I believe, been published. As it flows from the tree it is sweet, and void of intoxicating properties; but when allowed to stand for a short time it usually ferments, and becomes first intoxicating, and afterwards acid. Upon the tendency to ferment, the place of growth appears to have an influence. This is shown by the circumstance, that while the juice of the fan palm produces the usual toddy of India, that of the date tree is preferred to it among the hills of Bahar, because there the sap of the fan palm does not readily ferment—(HOOKER).

The date juice, in the Sahara, when drunk immediately, tastes like genuine rich milk; but when allowed to stand for a night, or at most for twenty-four hours, it ferments, and, except that it continues whitish, it acquires the sparkling quality and flavour of champagne. This quality no doubt differs with the kind of tree, and with the place of growth. By distillation the fermented juice yields a strong brandy, which is almost everywhere extracted from it in Africa, as well as in Asia. At Monghyr, on the banks of the Ganges—which is celebrated not only for its iron manufactures but for its drunkenness—Dr. Hooker observes that the abundance of toddy palms was quite remarkable.

In Chili, on the American coast, wine is made from a species of palm; in India, and other parts of Asia, palm wine is extensively consumed; while in Africa it is almost the only fermented liquor in very general use. Though we know so little of it in Europe, therefore, the wine of the palm tree is drunk as an exhilarating liquor by a larger number of the human race than the wine of the grape.

4°. SUGAR-CANE WINE, OR GUARAPO.—Like the sap of the palm tree, that of the sugar cane ferments spontaneously, and produces an intoxicating liquor. To this cane-wine the negroes give the name of Guarapo, and they hold it in high esteem. It contains, of course, all the ingredients of the cane juice, except those which are changed or naturally disappear during the fermentation, and those which subside when it clarifies. I am not aware, however, that any special chemical examination of this drink has hitherto been made.

5°. Pulque, Octli, or Agave Wine, is the favourite drink of the lower classes in the central part of the table-land of Mexico. It is produced by fermenting the sap of the Maguey or American aloe (*Agave Americana* or *Mexicana*), which is cultivated in plantations for the purpose. This plant is of slow growth, but when full grown its leaves attain a height of five to eight feet, and even more. It flowers on an average only once in ten years, and, as in the case of palm wine, it is from the flower-stalk that the juice is extracted. In the plantations, the Indian watches each plant as the time of its flowering approaches, and just when the central shoot or flower-stem is about to appear, he makes a deep cut, and scoops out the whole heart (*el corazon*) or middle part of the stem, leaving nothing but the outside rind. This forms a natural basin or well, about two feet in depth and one and a half in width. Into this well the sap, which was intended to feed the shoot, flows so rapidly that it is necessary to remove it twice, and sometime three times a-day. To make this more easy, the leaves on one side are cut away and the central basin laid open, as is seen in fig. 52.

The sap as it flows has a very sweet taste, and none of that disagreeable smell which it afterwards acquires. It is called *aguamiel* or honey-water. It ferments spontaneously, and a small quantity of old fermented juice speedily induces fermentation in that which is newly drawn, as sour leaven does in new dough. It is usual, therefore, to set aside a portion of sap, to ferment separately for ten or fifteen days, and to add a small quantity of this to each vessel of fresh juice. Fermentation is excited immediately, and

Fig. 52.

Agave Americana—The American Aloe.
As prepared for producing pulque, and with a distant flowering-plant.
Scale, 1 inch to 5 feet.

in twenty-four hours it becomes pulque in the very best state for drinking. A good maguey yields from eight to fifteen pints a-day, and this supply continues during two and often three months—(Ward).*

The chemical changes which take place during the fermentation of this juice are the more interesting as they are in some respects peculiar.

First, Alcohol is produced as in other fermented liquors. This is shown by the slightly intoxicating qualities of the drink, and by its yielding, when distilled, an ardent spirit. To this brandy the name of *mexical* is given, or of *aguardiente de maguey*. The average proportion of alcohol in the pulque is not stated.

* *Mexico in* 1827, vol. i., p. 57.

Second, An acid is formed also—the pulque, as a drink, being described as resembling cider. But what is the nature of the acid has not been determined. But,

Third, The most remarkable result of the fermentation is, that the nearly smell-less juice acquires a fetid and disagreeable odour of tainted meat. This makes the liquor be looked upon at first with disgust, especially by Europeans. It is so cool, agreeable, and refreshing, however, that this first disgust being overcome, the pulque is preferred, even by Europeans, to every other liquid.

The nature of this evil-smelling ingredient, and the chemical changes by which it is produced, have not been investigated. It is probably similar in kind to that which gives the bad smell to putrid fish (*Trimethylamine*).* Substances of this kind are sometimes produced in the living plant. The Bladder-headed Saussurea, for example, which grows in the Himalayas, emits as it grows the smell of putrid meat; and the *Stapelias* are called carrion-flowers, because of the disagreeable putrid odours they exhale.

The natives of Mexico ascribe many good qualities to their national drink. It is an excellent stomachic, promotes digestion, induces sleep, and is esteemed as a remedy in many diseases. It is chiefly in the neighbourhood of large towns, like Puebla and Mexico, that the maguey plantations exist. The pulque so soon passes that state of fermentation at which it is most pleasant to drink, that the manufacture only pays where a speedy sale is certain. The brandy or aguardiente, which is not liable to this inconvenience, is largely manufactured, and more widely consumed than the pulque itself.

* See in a subsequent chapter THE SMELLS WE DISLIKE.

CHAPTER XIV.

THE LIQUORS WE FERMENT.

THE BRANDIES.

The brandies, or ardent spirits.—Methods of distillation.—Absolute alcohol.—Strength of different varieties of spirits.—Peculiarities in the preparatory processes of the distiller.—Use of raw grain mixed with malt; profit of this.—Average produce of proof spirits.—Peculiar flavour of cognac, rum, &c.—Consumption of home-made ardent spirits in the three kingdoms.—Quantity of malt used in brewing.—Spirits consumed in the form of beer.—Comparative sobriety of England, Scotland, and Ireland.—Consumption of foreign liquors.—Alleged greater intemperance of Scotland and Ireland: how this impression has been produced.—Influence of the nutritive matter, and of the hops contained in beer.—Influence of general food and temperament.—Ardent spirits serve the same purpose as the starch and fat of our food, and retard the waste of the body.—Wine, "the milk of the aged."—Substances employed to give a fictitious strength to fermented liquors.

III. The Brandies, or Ardent Spirits.—When fermented liquors, such as those above described, are put into an open vessel and heated over a fire till they begin to boil, the alcohol they contain rises in the form of vapour, along with a little steam, and escapes into the air. If this boiling be performed in a close vessel, from which the vapours as they rise are conducted by a pipe into a cooled receiver, they condense again into a liquid state. This is the process called distillation, and the vessel in which it is carried on is called a still.

1°. The distillation.—A retort connected with a receiver, over which a stream of cold water is kept flowing (fig. 53), represents the simplest form of such a still; but

Fig. 53.

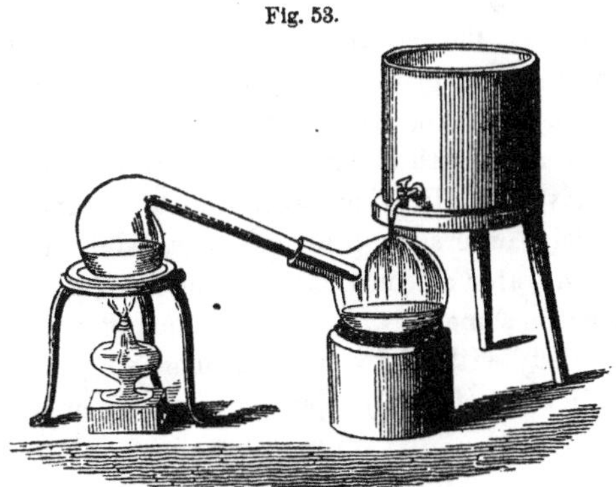

many more complicated forms of apparatus have been contrived for the purpose of conducting the process with economy and efficiency. The following illustration (fig. 54) represents

Fig. 54.

a form of still, of common use in our laboratories, for distilling water. The kettle A, which contains the water, is covered by the movable dome B, from which the pipe *b c* conducts the vapour into the receiver R, which is surrounded with cold water. Thence the condensed liquid descends through a continuation of the tube, bent spirally, called the *worm*, by which it is exposed to the prolonged action of the cold water, till at length it flows quite cool into the bottle placed to receive it. Into the worm-tub a stream of cold water constantly enters by the pipe *p p*, while a similar stream of warm water as constantly escapes by the pipe *q*.

Arrangements somewhat different are made in the large distilleries, chiefly with the view of economising time and fuel. The following (fig. 55) represents a common form of

Fig. 55.

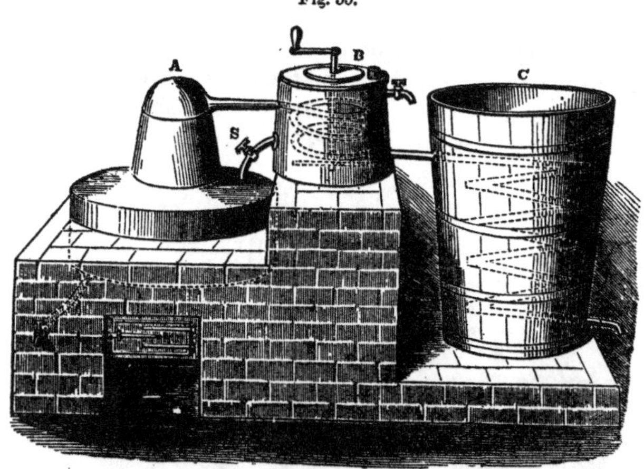

apparatus, where the process of spirit-distillation is conducted on a large scale. The principal peculiarities in this are —*first*, The broad flat bottom of the pot or still A, by which the effect of the heat is more quickly and fully obtained;

and, *second*, The adoption of two worms, B and C, in different vessels. In the first of these vessels cold wort is put, which is heated by the vapours as the distillation proceeds, and when hot is run at once by the stopcock s into the still. The second vessel contains cold water as before, and as this water heats it is run off, and is employed in mashing the grain. Thus heat is economised in various ways.

The spirit which passes off and condenses in the worm is more or less mixed with water, but by means of successive distillations—or *rectifications*, as they are called—it may be obtained quite free from water. It is then what chemists call absolute alcohol. This pure or absolute alcohol has a peculiar penetrating smell; a hot, fiery, and burning taste; is about one-fifth part lighter than water;* burns readily, but with a pale flame when kindled in the air, and is intoxicating in a high degree. It is used only for chemical purposes. The spirit-of-wine, or common alcohol of the shops, which we burn in our lamps, and employ for other familiar uses, is already diluted with a considerable proportion of water.

In the brandies, or varieties of ardent spirits which we consume as exhilarating drinks, the alcohol is still further diluted with water.

Thus the proportions of alcohol per cent., in some of the common varieties of commercial spirits, are as follows (at 62° Fahr.):—

	Alcohol.	
	By weight.	By measure.
British proof-spirit contains	50	57
Commercial Cognac, ...		50 to 54
Rum, ...		72 to 77
Geneva, ...		50
Whisky, ...		59

* A vessel which will hold 1000 grains of water will hold only 792 of absolute alcohol. Its specific gravity is therefore said to be 792, that of water being 1000—or 0.792, that of water being 1.

So that on an average, we may say that the ardent spirits we consume contain only half their weight, or three-fifths of their bulk of absolute alcohol. They are about twice as strong as our port, sherry, and Madeira wines.

Every different fermented liquor, when distilled, yields an ardent spirit which has a flavour, and is generally distinguished by a name of its own. Thus wine yields what we call brandy or cognac: fermented molasses yields rum; Indian corn, potatoes, and rye, yield liquors which are distinguished as corn, rye, and potato brandies; while malt liquors give our Scotch and Irish whiskies. If juniper berries be added previous to distillation, as is usually done in Holland, a flavour is imparted to the spirit which is characteristic of gin or Hollands; and if the malt be dried over a peat fire, the smell and taste of the peat (the peet-reek) accompany the spirit prepared from it; and these, in the estimation of the initiated, impart a peculiar value to peet-reek whisky.

2°. THE DISTILLERS' PROCESSES.—But though malt and other liquors, fermented in the usual way—indeed, in almost any way—will yield brandy by distillation, yet the distiller by profession conducts his fermenting operations in a somewhat different way from the brewer, whose object is merely the production of beer. Thus—

First, We have seen that, in fermenting the wort for the manufacture of beer, a large proportion of the sugar is left in the liquor unchanged. The fermentation is stopped before this sugar is transformed into alcohol, in order that the beer may be pleasant to drink, and that it may keep in the cask without turning sour. But the distiller's object is to obtain the largest possible quantity of spirit from his grain; he therefore prolongs the fermentation until the whole of the sugar is transformed, as nearly as possible, into alcohol and carbonic acid. To leave any of it unchanged would not only involve a loss of spirit, but, during the subsequent distillation, might injure the flavour and general quality of the spirit he obtained. The securing of this point, therefore, requires on his part an attention to minute circumstances, different a little in kind, but not less nice and delicate than those which determine the success of the brewer's operations.

Again, the most agreeable and generally esteemed grain-spirit is obtained when malted barley only is employed in the manufacture. This yields in Scotland and Ireland the best malt whisky. The profit of the distiller, however, is often promoted by mixing with the malt a greater or less proportion of unmalted grain, or even of potato starch. To the reason of this I have already briefly alluded (p. 243), but it is worthy of a fuller explanation.

We have seen that it is the diastase, produced during the germination of the barley, which subsequently transforms the starch of the grain into sugar. This diastase is capable of so transforming nearly a thousand times its own weight of starch; but good malt contains only a hundred of starch to one of diastase. The latter ingredient, therefore, will transform into sugar ten times as much starch as it is associated with in the best malt. Hence a large quantity of starch, either in the form of crushed unmalted grain, or of potato starch, may be mixed even with ordinary malt in the mash-tub, with the certainty that the diastase of the malt will transform it all into sugar.

This is what the distiller does in making *grain* whisky; and the profit of it consists in this—that he saves both the expense of malting his grain and the loss of matter (usually 8 per cent.),* which barley always undergoes in malting.

* A hundred pounds of barley yield only eighty pounds of malt. But of this loss 12 per cent. consists of water driven off by the heat of the malt kiln, so that the real loss of substance is 8 lb. in the 100.

He is able, also, to use for these additions of grain an inferior or cheaper material than is usually employed for conversion into malt.* The sweet wort obtained in this way, when fermented and distilled, yields a spirit of a somewhat harsher and less pleasant flavour than when malt alone is used.

Along with the spirit, during the distillation of fermented liquors, there always passes over a small but variable proportion of one or more volatile oily liquids, which mix with the spirit and give it a peculiar flavour. These volatile oils vary in kind, in composition, and in sensible properties, with the source of the sugar which has been submitted to fermentation, and with the substances which are present along with it in the wort. Hence the spirit obtained from almost every different fermented liquor is distinguished by its own characteristic flavour. Thus wine, brandy, or cognac, derives its vinous flavour from the juice of the grape; and cognacs of different districts their special flavours from the kinds of wine which are distilled in each. Rum obtains its smell and taste from molasses, the scorched and altered juice of the sugar cane; whisky its peculiarities from the barley-malt or grain that is mixed with it; potato brandy, from the mashed potato or its skin;† palm brandy,

* Thus, in some of the Scotch distilleries, such a mixture as the following is employed:—

Malt,	42	bushels at	40 lb.	a bushel.
Oats,	25	„	47	„
Rye,	25	„	53	„
Barley,	158	„	53	„
	250			

The diastase in the 42 bushels of malt converts into sugar the starch of the whole 250 bushels, weighing eight times as much as the malt itself. This quantity of grain yields on an average 583 gallons of proof whisky, or 14 gallons from 6 bushels of the mixture.

† Potato brandy is contaminated, among other substances, by a volatile spirit called *amyle alcohol*. And it is a singular circumstance that the cognac distilled in the south of France from the grape husks—known as Eau de vie de marc de raisin—also contains the same amyle alcohol. In the one case it is probably derived from the skin of the root, in the other from the skin of the fruit.

from the fermented toddy; the aguardiente of Mexico, from the strong-smelling pulque; and the arraca of the Kalmucks, from their fermented milk. And so with other varieties of spirit. In each case a volatile substance, peculiar in kind, accompanies the spirit; and though this substance is always very small in quantity, it is yet sufficient to impart to each different variety a flavour at once characteristic and peculiar to itself.

It is chiefly from malted and raw grain of various kinds that ardent spirits are distilled in the British islands, in Northern Europe generally, and in the North American states and colonies. Maize or Indian corn is most extensively employed for this purpose in the United States. Potatoes are used to a considerable extent on the continent of Europe; and sugar is occasionally employed in our own distilleries.

3°. CONSUMPTION OF ARDENT SPIRITS—The manufacture and consumption of ardent spirits, especially in northern climates, is exceedingly great. In the United Kingdom, the quantity distilled and consumed, in the year ending on the 5th of January, 1854, was about 25 millions of gallons, distributed as follows:—

	Distilled.	Consumed.
England, ...	10,729,243 gallons.	10,850,807 gallons.
Scotland, . . .	6,557,839 "	6,534,648 "
Ireland, ...	8,136,862 "	8,136,862 "
United Kingdom,	25,423,444	25,021,817

This is a very large quantity of ardent spirits to be consumed by a population of less than thirty millions. The numbers appear especially large in the cases of Scotland and Ireland, and would seem at first sight to imply a much greater proportionate consumption of alcohol in these countries than in England.

But a simple application of chemical knowledge materially alters this first conclusion.

a. In the year ending on the 10th October, 1852,* the quantity of malt consumed in each of the three kingdoms *in the making of beer*, was in bushels—

England, ...	30,636,240
Scotland, ...	1,127,224
Ireland, ...	1,266,344
United Kingdom,	33,029,808

From which numbers it appears, that of the 33 millions of bushels of malt used in the three kingdoms for the making of beer, 30½ millions are consumed in England alone.

Now, in the average of years, one bushel of malt yields two gallons of proof spirit, so that the *malt yearly made into beer in England, if employed for making whisky, would yield the enormous quantity of* 61 *millions of gallons!*

I have already stated, however, that in the fermentation of the worts for the manufacture of beer, the whole of the sugar is not transformed into alcohol. From one-fourth to sometimes one-half of the whole sugar remains unchanged in the beer. The quantity of malt, therefore, which is consumed in England for the making of this milder drink does not in reality indicate the consumption of so large a number of gallons of ardent spirits as the distiller would extract from it. If we allow one-fourth of the whole for the sugar remaining unchanged in the beer, then the quantity of ardent spirits actually consumed in the three kingdoms would be very nearly as follows (in gallons):—

* I use this return because I have not at hand any later one, which distinguishes the malt used by the brewers from that used by the distillers.

	England.	Scotland.	Ireland.
Spirits consumed as such,	10,350,307	6,534,648	8,136,362
Spirits consumed in the beer,	45,954,360	1,790,836	1,899,516
Total spirits consumed,	56,304,667	8,325,484	10,035,878

Now, if we divide these several total sums by the population of each of the three kingdoms, we obtain the following numbers for the quantity of ardent spirits consumed per head in each country—

	England.	Scotland.	Ireland.
Total consumption in gallons,	56⅓ millions.	8⅓ millions.	10 millions.
Population, ...	18 "	3 "	6½ "
Consumption per head in gallons,	$8\frac{1}{9}$ "	$2\frac{7}{9}$ "	1½ "

In so far as the mere consumption of alcohol, in the form of home-made liquors, goes, therefore, it appears that Scotland does not in reality surpass England. On the contrary, England somewhat exceeds Scotland, while both England and Scotland greatly surpass Ireland. For every head of its population, Ireland consumes less than half what is consumed in England, and somewhat more than half of what is consumed in Scotland. This very small comparative consumption in Ireland is not to be ascribed to an increased temperance caused by the labours of Father Matthew and others. On the contrary, since his time the consumption per head has greatly increased, as is seen by comparing the last two decennial periods. Thus—

In the year	The population was	And the consumption of spirits. Total.	Per head.
1842	8,175,124	5,299,650	5¼ pints.
1852	6,515,794	8,208,256	10 pints.

The consumption per head in Ireland is, therefore, rapidly increasing; and it is both fairer and safer, I think, to ascribe this increase to a general advance in material prosperity, than to augmenting intemperance and dissipation.

b. But in estimating the actual and relative consumption of alcohol in England and Scotland, there are still two other items to be taken into calculation. Wine and foreign spirits are imported into the United Kingdom, and consumed in large quantities. Thus, in the year ending 5th January, 1854 there was entered for home consumption, in gallons,—

	Gallons.	Containing of proof spirits. Gallons.
Wine,	7,197,572	1,440,000*
Foreign spirits,		5,131,618
	Total,	6,571,618

Now, in England, the consumption of wine and foreign spirits, among the middle and higher classes, is certainly far more universal than among the same classes in Scotland. A much larger proportion per head of the $6\frac{1}{2}$ millions of gallons of spirits, consumed in the form of imported liquors, must therefore be ascribed to England. Let us suppose it all to be consumed in Great Britain—leaving the small consumption of Ireland out of the question—and that every Englishman drinks two bottles for the Scotchman's one; then—

The Englishman drinks	$2\frac{2}{3}$ pints, and
The Scotchman	$1\frac{1}{3}$ pints

of ardent spirits, in the form of imported liquors. Adding this to the consumption, in the form of home-made liquors, we have the total consumption per head as follows, in gallons :—

* Supposing foreign wines to contain an average of only ten per cent. of alcohol, which is probably one-half too low.

	England.	Scotland.
In home-made liquors, . .	$3\frac{1}{9}$	$2\frac{14}{18}$
In imported liquors, . . .	$0\frac{3}{9}$	$0\frac{3}{18}$
Total per head. . . .	$3\frac{4}{9}$	$2\frac{17}{18}$

Or, in England, the total consumption is about $3\frac{1}{2}$, and in Scotland about 3 gallons per head. These numbers do not, in themselves, imply very extreme intemperance in either country. Were the total quantity of ardent spirits we use really equally distributed and consumed in the above proportions by the whole population, cases of drunkenness would not necessarily occur. It is because many consume more than their share that the evils of intemperance so often manifest themselves.

c. Two chemico-physiological points in connection with this subject are deserving of our consideration. It is very generally believed, and has recently at least been very often asserted—and what is curious, most strongly and earnestly in Scotland itself—that in Scotland intemperance is a much more common vice than in England. But how can this be, since the average individual consumption of alcohol in England is one-sixth part greater than in Scotland?

And, again, Ireland has been reproached for its intemperance and for its love of whisky even more than Scotland, and yet the individual consumption of alcohol in any form is probably less in that island than in any northern country, either European or American. Can this allegation be true, or how is it to be accounted for?

First, As to the alleged greater sobriety of England, it is to be observed, that upwards of three-fourths of all the alcohol drunk in that country is in the form of beer. This liquor, as we have seen, feeds and nourishes while it exhilarates the Englishman. All which the distillers' fermented wort contains, except its alcohol, remains behind in the still and is lost as food for man. All that the brewers' wort contains, with the exception of what separates in the fining of his liquor, is retained and drunk in the beer. Sugar and gluten to the amount of from 4 to 8 per cent. of its weight, exist in the malt liquor; and these, by strengthening the system, modify and mollify the apparent action of the alcohol with which they are associated. They place malt liquors in the same relation to ardent spirits as cocoa bears to tea and coffee.*

Besides, beer is drugged, so to speak, with hops, the tonic, narcotic, and sedative influences of which restrain, retard, and modify the intoxicating action of the spirit. Thus—controlled by the nutritive and narcotic ingredients it is associated with—a larger proportion of ardent spirit will produce a smaller sensible intoxicating effect than if taken alone. And thus, a people may appear more temperate and sober while in reality it consumes a larger proportion of ardent spirits.

Second.—But though these reasons may go far to explain the difference in the reputed sobriety of the two ends of our own island, they scarcely explain why Ireland, which consumes so little per head, should be charged with an amount of intemperance greater even than Scotland itself. Here I believe other causes come into play. Of these I instance only two—the less substantial food, and the more excitable temperament of the Irish people. Every one knows how easily a man becomes intoxicated if he pours down ardent spirits into an empty stomach. And from this extreme case the effect of a given quantity of spirits becomes less as the quantity of good food eaten becomes greater. It is least of all on the well-fed muscular beef-eating labourer.

* See THE BEVERAGES WE INFUSE.

And, again, excitable people, even when well fed, are influenced more than others by intoxicating drinks. As a people, it will, I believe, be conceded that the Irish are more excitable than the British; and likely, therefore, to be overcome by a quantity of liquor which persons of a more immovable temperament could, in the same circumstances, drink with impunity. It is probable that the quality and quantity of the national food has a material influence upon national temperament. But however this be, I am inclined to see, in the two things—in the national food and the national temperament—an explanation of the alleged insobriety of a people who, it is certain, do really consume so little intoxicating drink.*

This influence of temperament, in connection with that of climate, has probable something to do also with the great evils which are said to arise from the use of ardent spirits among the European races settled in North America. These, as is well known, have of late years given rise to much discussion—to strenuous efforts, on the part of the benevolent, to check the consumption of fermented liquors—and to the passing of what is called the Maine Law, for the purpose of effectually repressing it.

4°. Influence of ardent spirits.—In the ardour of this crusade against fermented liquors, statements have been hastily made by over-zealous champions of total abstinence, which are not quite borne out by chemical and physiological researches.

Ardent spirits of every variety are little else than alcohol diluted with a large proportion of water, and flavour-

* Good fellowship is an enemy to sobriety—not for the vulgar reason that it provokes to the passing of the bottle, but because it makes what is drunk have a greater apparent effect. It is familiar to the knowing ones, that if a man wishes to drink, he had better let his companions *do all the talking*. "Gin' ye're gaun to drink, sir, dinna ye talk muckle." Here the temperament of the mercurial and excitable tells at once.

ed with a minute admixture of volatile oil, the precise action of which upon the system is not known. They contain none, therefore, of the common forms of nutritive matter which exist in our usual varieties of animal and vegetable food. It does not follow from this, however, as some have too broadly alleged, that they are incapable of serving any useful purpose in the animal economy. On the contrary, it is ascertained of ardent spirits—

First, That they directly warm the body, and, by the changes they undergo in the blood, supply a portion of that carbonic acid and watery vapour which, as a necessity of life, are constantly being given off by the lungs. They so far, therefore, supply the place of food—of the fat and starch for example—which we usually eat. Hence a schnapps, in Germany, with a slice of lean dried meat, make a mixture like that of the starch and gluten in our bread, which is capable of feeding the body. So we either add sugar to milk, or take spirits along with it (old man's milk), for the purpose of adjusting the proportions of the ingredients more suitably to the constitution, or to the circumstances in which it is to be consumed.

Second, That they diminish the absolute amount of matter usually given off by the lungs and the kidneys. They thus lessen, as tea and coffee do, (p. 191,) the natural waste of the fat and tissues, and they necessarily diminish, in an equal degree, the quantity of ordinary food which is necessary to keep up the weight of the body. In other words, they have the property of making a given weight of food go further in sustaining the strength and bulk of the body. And in addition to the saving of material thus effected, they ease and lighten the labour of the digestive organs, which, when the stomach is weak, is often a most valuable result.

Hence fermented liquors, if otherwise suitable to the

constitution, exercise a beneficial influence upon old people, and other weakly persons whose fat and tissues have begun to waste—in whom the process of digestion, that is, does not replace the tissues as fast as they naturally waste. This lessening in weight or substance is one of the most usual consequences of the approach of old age. It is a common symptom of the decline of life. The stomach either does not receive or does not digest food enough to replace that which is daily removed from the substance of the body. Weak alcoholic drinks arrest or retard, and thus diminish the daily amount of this loss of substance. They gently stimulate the digestive organs also, and help them to do their work more fully and faithfully; and thus the body is sustained to a later period in life. Hence poets have called wine "the milk of the old," and scientific philosophy owns the propriety of the term. If it does not nourish the old so directly as milk nourishes the young, yet it certainly does aid in supporting and filling up their failing frames. And it is one of the happy consequences of a temperate youth and manhood, that this spirituous milk does not fail in its good effects when the weight of years begins to press upon us.

All this, of course, in no way justifies the indulgence in fermented liquors of any kind to excess, or palliates the moral evils to which this excess invariably gives rise. The good results I have spoken of follow only from a moderate use of them. But the peculiar danger attendant upon the consumption of intoxicating drinks arises from their extreme seductiveness, and from the all but unconquerable strength of the drinking habit when once formed. Their peculiar malignity appears—where they have once obtained a mastery—in their becoming the parent and nurse of every kind of suffering, immorality, and crime

"Who hath woe?" says Solomon; 'who hath sorrow?

who hath contentions? who hath babbling? who hath wounds without cause? who hath redness of eyes? They that tarry long at the wine; they that go to seek mixed wine. Look not thou upon the wine when it is red, when it giveth his colour in the cup, when it moveth itself aright (sparkleth?). At the last it biteth like a serpent, and stingeth like an adder."

5°. Adulteration of fermented liquors.—The real strength of pure fermented liquors depends, as we have seen, on the proportion of alcohol they contain. But in various countries adulterating substances are added to them, chiefly of a narcotic kind, for the purpose of imparting a fictitious or apparent strength.

Thus, to malt beer, *Cocculus indicus*, grains of paradise, the root of the sweet flag, and even tobacco-leaves, are added in England, the *Ledum palustre* and *Ledum latifolium*, in North Germany; the *Achillea millefolia*, or yarrow, in Dalecarlia; and the seeds of *Datura stramonium* in Russia, in India, and formerly in China. In Java, *ragi* cakes, made of onions, black pepper and capsicums, are fermented with boiled rice, to give a similar strength to rice beer.

To grape wine poppy heads are now added in Persia. In ancient Palestine, frankincense was added, especially to the wine given to criminals, for the purpose of stupefying them before the execution began; and in ancient Greece, sea-water in the proportion of 1 of water to 50 of wine, with the view of aiding digestion, and preventing its affecting the head.

To ardent spirits, seeds of thorn-apple are added in India; and in England, Malagueta pepper with capsicum, calamus, and juniper berries, to give a hot, strong flavour to London gin.

These substances are all foreign to the true nature and

composition of the liquors we ferment. They add nothing to the amount of alcohol contained in these liquors. They affect their quality generally by introducing narcotic ingredients. The chemical properties of most of these narcotic ingredients, and their action upon the system, will be treated of in the immediately succeeding chapters upon the NARCOTICS WE INDULGE IN.

END OF VOL. I

A LIST
OF
NEW WORKS
IN GENERAL LITERATURE,
PUBLISHED BY
D. APPLETON & COMPANY,
346 & 348 Broadway.

**** Complete Catalogues, containing full descriptions, to be had on application to the Publishers.*

Agriculture and Rural Affairs.

Boussingault's Rural Economy, ... 1 25
The Poultry Book, illustrated, ... 5 00
Waring's Elements of Agriculture, . . . 75

Arts, Manufactures, and Architecture.

Appleton's Dictionary of Mechanics. 2 vols. . 12 00
" Mechanics' Magazine. 3 vols. each, 3 50
Allen's Philosophy of Mechanics, ... 3 50
Arnot's Gothic Architecture, ... 4 00
Bassnett's Theory of Storms, ... 1 00
Bourne on the Steam Engine, ... 75
Byrne on Logarithms, ... 1 00
Chapman on the American Rifle, ... 1 25
Coming's Preservation of Health, ... 75
Cullum on Military Bridges, ... 2 00
Downing's Country Houses, ... 4 00
Field's City Architecture, ... 2 00
Griffith's Marine Architecture, ... 10 00
Gillespie's Treatise on Surveying, ...
Haupt's Theory of Bridge Construction, . . 3 00
Henck's Field-Book for R. Road Engineers, . 1 75
Hoblyn's Dictionary of Scientific Terms, . . 1 50
Huff's Manual of Electro-Physiology, . . . 1 25
Jeffers' Practice of Naval Gunnery, . . . 2 50
Knapen's Mechanics' Assistant, ... 1 00
Lafever's Modern Architecture, ... 4 0
Lyell's Manual of Geology, ... 1 75
" Principles of Geology, ... 2 25
Reynold's Treatise on Handrailing, . . . 2 00
Templeton's Mechanic's Companion, . . . 1 00
Ure's Dict'ry of Arts, Manufactures, &c. 2 vols. 5 00
Youmans' Class-Book of Chemistry, . . . 75
" Atlas of Chemistry. cloth, . . . 2 00
" Alcohol, ... 50

Biography.

Arnold's Life and Correspondence, ... 2 00
Capt. Canot, or Twenty Years of a Slaver, . 1 25
Cousin's De Longueville, ... 1 00
Croswell's Memoirs, ... 2 00
Evelyn's Life of Godolphin, ... 50
Garland's Life of Randolph, ... 1 50
Gilfillan's Gallery of Portraits. 2d Series, . 1 00
Hernan Cortez's Life, ... 38
Hull's Civil and Military Life, ... 2 00
Life and Adventures of Daniel Boone, . . . 38
Life of Henry Hudson, ... 38
Life of Capt. John Smith, ... 38
Moore's Life of George Castriot, ... 1 00
Napoleon's Memoirs. By Duchess D'Abrantes, 4 00
Napoleon. By Laurent L'Ardèche, . . . 3 00
Pinkney (W.) Life. By his Nephew, . . . 2 00
Party Leaders: Lives of Jefferson, &c. . . 1 00
Southey's Life of Oliver Cromwell, . . . 38
Wynne's Lives of Eminent Men, ... 1 00
Webster's Life and Memorials. 2 vols. . . 1 00

Books of General Utility.

Appletons' Southern and Western Guide, . 1 00
" Northern and Eastern Guide, . 1 25
Appletons' Complete U. S. Guide, . . . 2 00
" Map of N. Y. City, . . . 25
American Practical Cook Book, . . .
A Treatise on Artificial Fish-Breeding, . . 75
Chemistry of Common Life. 2 vols. 12mo. .
Cooley's Book of Useful Knowledge, . . . 25
Cust's Invalid's Own Book, ... 50
Delisser's Interest Tables, ... 4 00
The English Cyclopaedia, per vol. . . . 2 50
Miles on the Horse's Foot, ... 25
The Nursery Basket. A Book for Young Mothers, 38
Pell's Guide for the Young, ... 38
Reid's New English Dictionary, ... 1 00
Stewart's Stable Economy, ... 1 00
Spalding's Hist. of English Literature, . . 1 00
Soyer's Modern Cookery, ... 1 00
The Successful Merchant, ... 1 00
Thomson on Food of Animals, ... 50

Commerce and Mercantile Affairs.

Anderson's Mercantile Correspondence, . . 1 00
Delisser's Interest Tables, ... 4 00
Merchants' Reference Book, ... 4 00
Oates' (Geo.) Interest Tables at 6 Per Cent. per Annum. 8vo. 2 00
" " Do. do. Abridged edition, . 1 25
" " 7 Per Cent. Interest Tables, . 2 00
" " Abridged, ... 1 25
Smith's Mercantile Law, ... 4 00

Geography and Atlases.

Appleton's Modern Atlas. 34 Maps, . . . 3 50
" Complete Atlas. 61 Maps, . . 9 00
Atlas of the Middle Ages. By Kœppen, . . 4 50
Black's General Atlas. 71 Maps, . . . 12 00
Cornell's Primary Geography, ... 50
" Intermediate Geography, ...
" High School Geography, ...

History.

Arnold's History of Rome, ... 3 00
" Later Commonwealth, ... 2 50
" Lectures on Modern History, . . 1 25
Dew's Ancient and Modern History, . . . 2 00
Koeppen's History of the Middle Ages. 2 vols. 2 50
" The same, folio, with Maps, . . 4 50
Kohlrausch's History of Germany, . . . 1 50
Mahon's (Lord) History of England, 2 vols. . 4 00
Michelet's History of France, 2 vols. . . 3 50
" History of the Roman Republic, . 1 00
Rowan's History of the French Revolution, . 63
Sprague's History of the Florida War, . . 2 50
Taylor's Manual of Ancient History, . . . 1 25
" Manual of Modern History, . . . 1 50
" Manual of History. 1 vol. complete, 2 50
Thiers' French Revolution. 4 vols. Illustrated, 5 00

Illustrated Works for Presents.

Bryant's Poems. 16 Illustrations. 8vo. cloth, . 3 50
" " " cloth, gilt, . 4 50
" " " mor. antique, 6 00

Gems of British Art. 30 Engravings. 1 vol. 4to. morocco, ... 18 00
Gray's Elegy. Illustrated. 8vo. 1 50
Goldsmith's Deserted Village, . . . 1 50
The Homes of American Authors. With Illustrations, cloth, ... 4 00
" " " cloth, gilt, 5 00
" " " mor. antqe. 7 00
The Holy Gospels. With 40 Designs by Overbeck. 1 vol. folio. Antique mor. . . 20 00
The Land of Bondage. By J. M. Wainwright, D. D. Morocco, ... 6 00
The Queens of England. By Agnes Strickland. With 29 Portraits. Antique mor. . . 10 00
The Ornaments of Memory. With 18 Illustrations. 4to. cloth, gilt, ... 6 00
" " Morocco, . . 10 00
Royal Gems from the Galleries of Europe. 40 Engravings, ... 25 00
The Republican Court; or, American Society in the Days of Washington. 21 Portraits. Antique mor. 12 00
The Vernon Gallery. 67 Engrav'gs. 4to. Ant. 25 00
The Women of the Bible. With 18 Engravings. Mor. antique, ... 10 00
Wilkie Gallery. Containing 60 Splendid Engravings. 4to. Antique mor. 25 00
A Winter Wreath of Summer Flowers. By S. G. Goodrich. Illustrated. Cloth, gilt, . 3 00

Juvenile Books.

A Poetry Book for Children, ... 75
Aunt Fanny's Christmas Stories, . . . 50
American Historical Tales, ... 75

UNCLE AMEREL'S STORY BOOKS.

The Little Gift Book. 18mo. cloth, . . 25
The Child's Story Book. Illust. 18mo. cloth, 25
Summer Holidays. 18mo. cloth, . . . 25
Winter Holidays. Illustrated. 18mo. cloth, . 25
George's Adventures in the Country. Illustrated. 18mo. cloth, ... 25
Christmas Stories. Illustrated. 18mo. cloth, . 25

Book of Trades, ... 50
Boys at Home By the Author of Edgar Clifton, 75
Child's Cheerful Companion, ... 50
Child's Picture and Verse Book. 100 Engs. . 50

COUSIN ALICE'S WORKS.

All's Not Gold that Glitters, ... 75
Contentment Better than Wealth, . . . 63
Nothing Venture, Nothing Have, . . . 63
No such Word as Fail, ... 63
Patient Waiting No Loss, ... 63

Dashwood Priory. By the Author of Edgar Clifton, ... 75
Edgar Clifton; or Right and Wrong, . . 75
Fireside Fairies. By Susan Pindar, . . 63
Good in Every Thing. By Mrs. Barwell, . 50
Leisure Moments Improved, , ... 75
Life of Punchinello, ... 75

LIBRARY FOR MY YOUNG COUNTRYMEN.

Adventures of Capt. John Smith. By the Author of Uncle Philip, ... 38
Adventures of Daniel Boone. By do. . . 38
Dawnings of Genius. By Anne Pratt, . . 38
Life and Adventures of Henry Hudson. By the Author of Uncle Philip, ... 38
Life and Adventures of Hernan Cortez. By do. 38
Philip Randolph. A Tale of Virginia. By Mary Gertrude, ... 38
Rowan's History of the French Revolution. 2 vols. 75
Southey's Life of Oliver Cromwell, . . 38

Louis' School-Days. By E. J. May, . . . 75
Louise; or, The Beauty of Integrity, . . 25
Maryatt's Settlers in Canada, ... 62
" Masterman Ready, ... 63
" Scenes in Africa, ... 63
Midsummer Fays. By Susan Pindar, . . 63

MISS McINTOSH'S WORKS.

Aunt Kitty's Tales, 12mo. 75
Blind Alice; A Tale for Good Children, . . 38
Ellen Leslie; or, The Reward of Self-Control, 38
Florence Arnott; or, Is She Generous? . 38
Grace and Clara; or, Be Just as well as Generous, ... 38
Jessie Graham; or, Friends Dear, but Truth Dearer, ... 38
Emily Herbert; or, The Happy Home, . . 37
Rose and Lillie Stanhope, ... 37

Mamma's Story Book, ... 75
Pebbles from the Sea-Shore, ... 37
Puss in Boots. Illustrated. By Otto Specter, . 25

PETER PARLEY'S WORKS.

Faggots for the Fireside, ... 1 13
Parley's Present for all Seasons, . . . 1 00
Wanderers by Sea and Land, ... 1 13
Winter Wreath of Summer Flowers, . . 3 00

TALES FOR THE PEOPLE AND THEIR CHILDREN.

Alice Franklin. By Mary Howitt, ... 38
Crofton Boys (The). By Harriet Martineau, . 38
Dangers of Dining Out. By Mrs. Ellis, . . 38
Domestic Tales. By Hannah More. 2 vols. . 75
Early Friendship. By Mrs. Copley, . . 38
Farmer's Daughter (The). By Mrs. Cameron, 38
First Impressions. By Mrs. Ellis, . . . 38
Hope On, Hope Ever! By Mary Howitt, . 38
Little Coin, Much Care. By do. . . . 38
Looking-Glass for the Mind. Many plates, . 38
Love and Money. By Mary Howitt, . . 38
Minister's Family. By Mrs. Ellis, . . . 38
My Own Story. By Mary Howitt, . . . 38
My Uncle, the Clockmaker. By do. . . 38
No Sense Like Common Sense. By do. . 38
Peasant and the Prince. By H. Martineau, . 28
Poplar Grove. By Mrs. Copley, . . . 38
Somerville Hall. By Mrs. Ellis, . . . 38
Sowing and Reaping. By Mary Howitt, . 38
Story of a Genius. 38
Strive and Thrive. By do. 38
The Two Apprentices. By do. . . . 38
Tired of Housekeeping. By T. S. Arthur, . 38
Twin Sisters (The). By Mrs. Sandham, . 38
Which is the Wiser? By Mary Howitt, . 38
Who Shall be Greatest? By do. . . . 38
Work and Wages. By do. 38

SECOND SERIES.

Chances and Changes. By Charles Burdett, . 38
Goldmaker's Village. By H. Zschokke, . 38
Never Too Late. By Charles Burdett, . . 38
Ocean Work, Ancient and Modern. By J. H. Wright, ... 38

Picture Pleasure Book, 1st Series, . . . 1 25
" " " 2d Series, . . . 1 25
Robinson Crusoe. 300 Plates, ... 1 50
Susan Pindar's Story Book, ... 75
Sunshine of Greystone, ... 75
Travels of Bob the Squirrel, ... 37
Wonderful Story Book, ... 50
Willy's First Present, ... 75
Week's Delight; or, Games and Stories for the Parlor, ... 75
William Tell, the Hero of Switzerland, . . 50
Young Student. By Madame Guizot, . . 75

Miscellaneous and General Literature.

An Attic Philosopher in Paris, 25
Appletons' Library Manual, 1 25
Agnell's Book of Chess, 1 25
Arnold's Miscellaneous Works, 2 00
Arthur. The Successful Merchant, 75
A Book for Summer Time in the Country, 50
Baldwin's Flush Times in Alabama, 1 25
Calhoun (J. C.), Works of. 4 vols. publ., each, 2 00
Clark's (W. G.) Knick-Knacks, 1 25
Cornwall's Music as it Was, and as it Is, 63
Essays from the London Times. 1st & 2d Series, each, 50
Ewbanks' World in a Workshop, 75
Ellis' Women of England, 50
" Hearts and Homes, 1 50
" Prevention Better than Cure, 75
Foster's Essays on Christian Morals, 50
Goldsmith's Vicar of Wakefield, 75
Grant's Memoirs of an American Lady, 75
Gaieties and Gravities. By Horace Smith, 50
Guizot's History of Civilization, 1 00
Hearth-Stone. By Rev. S. Osgood, 1 00
Hobson. My Uncle and I, 75
Ingoldsby Legends, 50
Isham's Mud Cabin, 1 00
Johnson's Meaning of Words, 1 00
Kavanagh's Women of Christianity, 75
Leger's Animal Magnetism, 1 00
Life's Discipline. A Tale of Hungary, 63
Letters from Rome. A. D. 138, 1 90
Margaret Maitland, 75
Maiden and Married Life of Mary Powell, 50
Morton Montague; or a Young Christian's Choice, 75
Macaulay's Miscellanies. 5 vols. 5 00
Maxims of Washington. By J. F. Schroeder, 1 00
Mile Stones in our Life Journey, 1 00

MINIATURE CLASSICAL LIBRARY.

Poetic Lacon; or, Aphorisms from the Poets, 38
Bond's Golden Maxims, 31
Clarke's Scripture Promises. Complete, 38
Elizabeth; or, The Exiles of Siberia, 31
Goldsmith's Vicar of Wakefield, 38
" Essays, 38
Gems from American Poets, 38
Hannah More's Private Devotions, 31
" " Practical Piety. 2 vols. 75
Hemans' Domestic Affections, 31
Hoffman's Lays of the Hudson, &c. 38
Johnson's History of Rasselas, 38
Manual of Matrimony, 31
Moore's Lalla Rookh, 38
" Melodies. Complete, 38
Paul and Virginia, 31
Pollok's Course of Time, 38
Pure Gold from the Rivers of Wisdom, 38
Thomson's Seasons, 38
Token of the Heart. Do. of Affection. Do. of Remembrance. Do. of Friendship. Do. of Love. Each, 31
Useful Letter-Writer, 38
Wilson's Sacra Privata, 31
Young's Night Thoughts, 38

Little Pedlington and the Pedlingtonians, 50
Prismatics. Tales and Poems, 1 25
Papers from the Quarterly Review, 50
Republic of the United States. Its Duties, &c 1 00
Preservation of Health and Prevention of Disease, 75
School for Politics. By Chas. Gayerre, 75
Select Italian Comedies. Translated, 75
Shakespeare's Scholar. By R. G. White, 2 50
Spectator (The). New ed. 6 vols. cloth, 9 00
Swett's Treatise on Diseases of the Chest, 3 00
Stories from Blackwood, 50

THACKERAY'S WORKS.

The Book of Snobs, 50
Mr. Browne's Letters, 50
The Confessions of Fitzboodle, 50
The Fat Contributor, 50
Jeames' Diary. A Legend of the Rhine, 50
The Luck of Barry Lyndon, 1 00
Men's Wives, 50
The Paris Sketch Book. 2 vols. 1 00
The Shabby Genteel Story, 50
The Yellowplush Papers. 1 vol. 16mo. 50
Thackeray's Works. 6 vols. bound in cloth, 6 00

Trescott's Diplomacy of the Revolution, 75
Tuckerman's Artist Life, 75
Up Country Letters, 75
Ward's Letters from Three Continents, 1 00
" English Items, 1 00
Warner's Rudimental Lessons in Music, 50
Woman's Worth, 38

Philosophical Works.

Cousin's Course of Modern Philosophy, 3 00
" Philosophy of the Beautiful, 62
" on the True, Beautiful, and Good, 1 50
Comte's Positive Philosophy. 2 vols. 4 00
Hamilton's Philosophy. 1 vol. 8vo. 1 50

Poetry and the Drama.

Amelia's Poems. 1 vol. 12mo. 1 25
Brownell's Poems. 12mo. 75
Bryant's Poems. 1 vol. 8vo. Illustrated, 3 50
" " Antique mor. 6 00
" " 2 vols. 12mo. cloth, 2 00
" " 1 vol. 18mo. 63
Byron's Poetical Works. 1 vol. cloth, 3 00
" " " Antique mor. 6 00
Burns' Poetical Works. Cloth, 1 00
Butler's Hudibras. Cloth, 1 00
Campbell's Poetical Works. Cloth, 1 00
Coleridge's Poetical Works. Cloth, 1 25
Cowper's Poetical Works, 1 00
Chaucer's Canterbury Tales, 1 00
Dante's Poems. Cloth, 1 00
Dryden's Poetical Works. Cloth, 1 00
Fay (J. S.), Ulric; or, The Voices, 75
Goethe's Iphigenia in Tauris. Translated, 75
Gilfillan's Edition of the British Poets. 12 vols. published. Price per vol. cloth, 1 00
Do. do. Calf, per vol. 2 50
Griffith's (Mattie) Poems, 75
Hemans' Poetical Works. 2 vols. 16mo. 2 00
Herbert's Poetical Works. 16mo. cloth, 1 00
Keats' Poetical Works. Cloth, 12mo. 1 25
Kirke White's Poetical Works. Cloth, 1 00
Lord's Poems. 1 vol. 12mo. 75
" Christ in Hades. 12mo. 75
Milton's Paradise Lost. 18mo. 38
" Complete Poetical Works, 1 00
Moore's Poetical Works. 8vo. Illustrated, 3 00
" " " Mor. extra, 6 00
Montgomery's Sacred Poems. 1 vol. 12mo. 75
Pope's Poetical Works. 1 vol. 16mo. 1 00
Southey's Poetical Works. 1 vol. 3 00
Spenser's Faerie Queene. 1 vol. cloth, 1 00
Scott's Poetical Works. 1 vol. 1 00
" Lady of the Lake. 16mo. 38
" Marmion, 37
" Lay of the Last Minstrel, 25
Shakspeare's Dramatic Works, 2 00
Tasso's Jerusalem Delivered. 1 vol. 16mo. 1 00
Wordsworth (W.). The Prelude, 1 00

Religious Works.

Arnold's Rugby School Sermons, 50
Anthon's Catechism on the Homilies, 06
" Early Catechism for Children, 06
Burnet's History of the Reformation. 3 vols. 2 50
" Thirty-Nine Articles, 2 00

Bradley's Family and Parish Sermons,	2 00
Cotter's Mass and Rubrics,	38
Coit's Puritanism,	1 00
Evans' Rectory of Valehead,	50
Grayson's True Theory of Christianity,	1 00
Gresley on Preaching,	1 25
Griffin's Gospel its Own Advocate,	1 00
Hecker's Book of the Soul,	
Hooker's Complete Works. 2 vols.	4 00
James' Happiness,	25
James on the Nature of Evil,	1 00
Jarvis' Reply to Milner,	75
Kingsley's Sacred Choir,	75
Keble's Christian Year,	37
Layman's Letters to a Bishop,	25
Logan's Sermons and Expository Lectures,	1 13
Lyra Apostolica,	50
Marshall's Notes on Episcopacy,	1 00
Newman's Sermons and Subjects of the Day,	1 00
" Essay on Christian Doctrine,	75
Ogilby on Lay Baptism,	50
Pearson on the Creed,	2 00
Pulpit Cyclopædia and Ministers' Companion,	2 50
Sewell's Reading Preparatory to Confirmation,	75
Southard's Mystery of Godliness,	75
Sketches and Skeletons of Sermons,	2 50
Spencer's Christian Instructed,	1 00
Sherlock's Practical Christian,	75
Sutton's Disce Vivere--Learn to Live,	75
Swartz's Letters to my Godchild,	38
Trench's Notes on the Parables,	1 75
" Notes on the Miracles.	1 75
Taylor's Holy Living and Dying,	1 00
" Episcopacy Asserted and Maintained,	75
Tyng's Family Commentary,	2 00
Walker's Sermons on Practical Subjects,	2 00
Watson on Confirmation,	06
Wilberforce's Manual for Communicants,	38
Wilson's Lectures on Colossians,	75
Wyatt's Christian Altar,	38

Voyages and Travels.

Africa and the American Flag,	1 25
Appletons' Southern and Western Guide,	1 00
" Northern and Eastern Guide,	1 25
" Complete U. S. Guide Book,	2 00
" N. Y. City Map,	25
Bartlett's New Mexico, &c. 2 vols. Illustrated,	5 00
Burnet's N. Western Territory,	2 00
Bryant's What I Saw in California,	1 25
Coggeshall's Voyages. 2 vols.	2 50
Dix's Winter in Madeira,	1 00
Huc's Travels in Tartary and Thibet. 2 vols.	1 00
Layard's Nineveh. 1 vol. 8vo.	1 25
Notes of a Theological Student. 12mo.	1 00
Oliphant's Journey to Katmundu,	50
Parkyns' Abyssinia. 2 vols.	2 50
Russia as it Is. By Gurowski,	1 00
" By Count de Custine,	1 25
Squier's Nicaragua. 2 vols.	5 00
Tappan's Step from the New World to the Old,	1 75
Wanderings and Fortunes of Germ. Emigrants,	75
Williams' Isthmus of Tehuantepec. 2 vols. 8vo.	3 50

Works of Fiction.

GRACE AGUILAR'S WORKS.

The Days of Bruce. 2 vols. 12mo.	1 50
Home Scenes and Heart Studies. 12mo.	75
The Mother's Recompense. 12mo.	75
Woman's Friendship. 12mo.	75
Women of Israel. 2 vols. 12mo.	1 50

Basil. A Story of Modern Life. 12mo.	75
Brace's Fawn of the Pale Faces. 12mo.	75
Busy Moments of an Idle Woman,	75
Chestnut Wood. A Tale. 2 vols.	1 75
Don Quixotte, Translated. Illustrated,	1 25
Drury (A. H.). Light and Shade,	75
Dupuy (A. E.). The Conspirator,	75
Elten Parry; or, Trials of the Heart,	63

MRS. ELLIS' WORKS.

Hearts and Homes; or, Social Distinctions,	1 50
Prevention Better than Cure,	75
Women of England,	50

Emmanuel Phillibert. By Dumas,	1 25
Farmingdale. By Caroline Thomas,	1 00
Fullerton (Lady G.). Ellen Middleton,	75
" " Grantley Manor. 1 vol. 12mo.	75
" " Lady Bird. 1 vol. 12mo.	75
The Foresters. By Alex. Dumas,	75
Gore (Mrs.). The Dean's Daughter. 1 vol. 12mo.	75
Goldsmith's Vicar of Wakefield. 12mo.	75
Gil Blas. With 500 Engravings. Cloth, gt. edg.	2 50
Harry Muir. A Tale of Scottish Life,	75
Hearts Unveiled; or, I Knew You Would Like Him,	75
Heartsease; or, My Brother's Wife. 2 vols.	1 50
Heir of Redclyffe. 2 vols. cloth,	1 50
Heloise; or, The Unrevealed Secret. 12mo.	75
Hobson. My Uncle and I. 12mo.	75
Holmes' Tempest and Sunshine. 12mo.	1 00
Home is Home. A Domestic Story,	75
Howitt (Mary). The Heir of West Wayland,	50
Io. A Tale of the Ancient Fane. 12mo.	75
The Iron Cousin. By Mary Cowden Clarke,	1 25
James (G. P. R.). Adrian; or, Clouds of the Mind,	75
John; or, Is a Cousin in the Hand Worth Two in the Bush,	25

JULIA KAVANAGH'S WORKS.

Nathalie. A Tale. 12mo.	1 00
Madeline. 12mo.	75
Daisy Burns. 12mo.	1 00

Life's Discipline. A Tale of Hungary,	63
Lone Dove (The). A Legend,	75
Linny Lockwood. By Catherine Crowe,	50

MISS McINTOSH'S WORKS.

Two Lives; or, To Seem and To Be. 12mo.	75
Aunt Kitty's Tales. 12mo.	75
Charms and Counter-Charms. 12mo.	1 00
Evenings at Donaldson Manor,	75
The Lofty and the Lowly. 2 vols.	1 50

Margaret's Home. By Cousin Alice,	
Marie Louise; or, The Opposite Neighbors,	50
Maiden Aunt (The). A Story,	75
Manzoni. The Betrothed Lovers. 2 vols.	1 50
Margaret Cecil; or, I Can Because I Ought,	75
Morton Montague; or, The Christian's Choice,	75
Norman Leslie. By G. C. H.	75
Prismatics. Tales and Poems. By Haywarde,	1 25
Roe (A. S.). James Montjoy. 12mo.	75
" To Love and to Be Loved. 12mo.	75
" Time and Tide. 12mo.	75
Reuben Medlicott; or, The Coming Man,	75
Rose Douglass. By S. R. W.	75

MISS SEWELL'S WORKS.

Amy Herbert. A Tale. 12mo.	75
Experience of Life. 12mo.	75
Gertrude. A Tale. 12mo.	75
Katherine Ashton. 2 vols. 12mo.	1 50
Laneton Parsonage. A Tale. 3 vols. 12mo.	2 25
Margaret Percival. 2 vols.	1 50
Walter Lorimer, and Other Tales. 12mo.	75
A Journal Kept for Children of a Village School,	1 00

Sunbeams and Shadows. Cloth,	75
Thorpe's Hive of the Bee Hunter,	1 00
Thackeray's Works. 6 vols. 12mo.	6 00
The Virginia Comedians. 2 vols. 12mo.	1 50
Use of Sunshine. By S. M. 12mo.	75
Wight's Romance of Abelard & Heloise. 12mo.	75

The most Authentic and Entertaining Life of Napoleon.

Memoirs of Napoleon,

HIS COURT AND FAMILY.

BY THE DUCHESS D'ABRANTES, (MADAME JUNOT.)

Two Volumes, 8vo. 1134 pages. Price $4.

List of Steel Engravings contained in this Illustrated Edition.

NAPOLEON.
JOSEPHINE.
MARIA LOUISA,
DUKE OF REICHSTADT,
MADAME LAETITIA BONAPARTE,
CHARLES BONAPARTE,
LUCIEN BONAPARTE,
MARSHAL JUNOT,
CHARLES BONAPARTE,
PAULINE BONAPARTE,
ELIZA BONAPARTE,
JEROME BONAPARTE,
LOUIS BONAPARTE,
CARDINAL FESCH,
LOUISA, QUEEN OF PRUSSIA,
JOSEPH BONAPARTE.

Probably no writer has had the same opportunities for becoming acquainted with

NAPOLEON THE GREAT

as the Duchess D'Abrantes. Her mother rocked him in his cradle, and when he quitted Brienne and came to Paris, she guided and protected his younger days. Scarcely a day passed without his visiting her house during the period which preceded his departure for Italy as

COMMANDER-IN-CHIEF.

Abundant occasion was therefore had for watching the development of the great genius who afterwards became the master of the greater part of Europe.

MARSHAL JUNOT,

who became allied to the author of this work by marriage, was the intimate friend of Napoleon, and figured in most of the

BRILLIANT ENGAGEMENTS

which rendered him the greatest military captain of the age. No interruption took place in the intimacy which she enjoyed, so that in all these scenes, embracing a period of nearly

THIRTY YEARS,

the Duchess became familiar with all the secret springs of

NAPOLEON'S ACTIONS,

either through her husband or by her own personal knowledge and observation at the Court of Napoleon.

JOSEPHINE,

whose life and character so peculiarly attract the attention of all readers, occupies a great part of the first volume. The character and the deeds of

THE EMPERORS AND KINGS,
THE GREAT MEN OF THE DAY,
THE MARSHALS OF THE EMPIRE,
THE DISTINGUISHED LADIES OF THE COURT,

are described with minuteness, which personal observation only admits of. The work is written in that

FAMILIAR GOSSIPING STYLE,

and so interspersed with anecdotes that the reader never wearies. She has put every thing in her book—great events and small.

BATTLES AND BALLS,
COURT INTRIGUES AND BOUDOIR GOSSIP,
TREATIES AND FLIRTATIONS,

making two of the most charming volumes of memoirs, which will interest the reader in spite of himself.

Opinions of the Press.

"These anecdotes of Napoleon are the best yet given to the world, because the most intimate and familiar."—*London Literary Gazette.*

"We consider the performance now before us as more authentic and amusing than any other of its kind."—*London Quarterly Review.*

"Every thing relating to Napoleon is eagerly sought for and read in this country as well as in Europe, and this work, with its extraordinary attractions, will not fail to command a wide circulation. Madame Junot possessed qualifications for writing a semi-domestic history of the great Corsican which no other person, male or female, could command."—*Life Illustrated.*

A Work abounding in Exciting Scenes and Remarkable Incidents.

Capt. Canot;

OR,

TWENTY YEARS OF AN AFRICAN SLAVER:

BEING AN ACCOUNT OF HIS CAREER AND ADVENTURES ON THE COAST, IN THE INTERIOR, ON SHIPBOARD, AND IN THE WEST INDIES.

Written out and Edited from the Captain's Journals, Memoranda, and Conversations.

BY BRANTZ MAYER.

One Volume, 12mo. With eight Illustrations. Price $1 25.

Criticisms of the Press.

"The author is a literary gentleman of Baltimore, no Abolitionist, and we believe the work to be a truthful account of the life of a man who saw much more than falls to the lot of most men."—*Commonwealth.*

"A remarkable volume is this; because of its undoubted truth: it having been derived by Mayer from personal conversations with Canot, and from journals which the slaver furnished of his own life."—*Worcester Palladium.*

"Capt. Canot, the hero of the narrative, is, to our own knowledge, a veritable personage, and resides in Baltimore. There is no doubt that the main incidents connected with his extraordinary career are in every respect true."—*Arthur's Home Gazette.*

"Under one aspect, as the biography of a remarkable man who passed through a singularly strange and eventful experience, it is as interesting as any sea story that we have ever read."—*Boston Evening Traveller.*

"Capt. Canot has certainly passed through a life of difficulty, danger, and wild, daring adventure, which has much the air of romance, and still he, or rather his editor, tells the tale with so much straightforwardness, that we cannot doubt its truthfulness."—*New York Sunday Despatch.*

"The work could not have been better done if the principal actor had combined the descriptive talent of De Foe with the astuteness of Fouche and the dexterity of Gi. Blas, which traits are ascribed to the worthy whose acquaintance we shall soon make by his admiring editor."—*N. Y. Tribune.*

"The general style of the work is attractive, and the narrative spirited and bold—well suited to the daring and hazardous course of life led by the adventurer. This book is illustrated by several excellent engravings."—*Baltimore American.*

"The biography of an African slaver as taken from his own lips, and giving his adventures in this traffic for twenty years. With great natural keenness of perception and complete communicativeness, he has literally unmasked his real life, and tells both what he was and *what he saw*, the latter being the *Photograph* of the Negro in Africa, which has been so long wanted. A nephew of Mr. Mayer has illustrated the volume with eight admirable drawings. We should think no book of the present day would be received with so keen an interest."—*Home Journal.*

"Capt. Canot has passed most of his life since 1819 on the ocean, and his catalogue of adventures at sea and on land, rival in grotesqueness and apparent improbability the marvels of Robinson Crusoe."—*Evening Post.*

"If stirring incidents, hair-breadth escapes, and variety of adventure, can make a book interesting, this must possess abundant attractions."—*Newark Daily Advertiser.*

"This is a true record of the life of one who had spent the greater part of his days in dealing in human flesh. We commend this book to all lovers of adventure."—*Boston Christian Recorder.*

"We would advise every one who is a lover of 'books that are books'—every one who admires Le Sage and De Foe, and has lingered long over the charming pages of Gil Blas and Robinson Crusoe—every one, pro-slavery or anti-slavery, to purchase this book."—*Buffalo Courier.*

"Chestnut Wood will light up many a hearth with pleasure."

CHESTNUT WOOD:

An American Tale.

BY LIELE LINDEN.

Two volumes, 12mo. Paper covers, $1 25; cloth, $1 75.

PLOT OF THE STORY.

Chestnut Wood is a country-seat, near Sleepy Hollow, owned and occupied by Mr. Atherton, a man of stern but not unkind disposition. The better feelings of his heart are brought into action, by the circumstances of his young grand-daughter, Sybil, the heroine of the tale, who is thrown, by the death of her mother at a farm-house in the vicinity, where she has been rescued from exposure on the road, upon his protection. The father of Sybil, as may be inferred from the fate of her mother, is a worthless scoundrel, who endeavors, with the help of associates as worthless as himself, to get possession of the child. They succeed in carrying her off, and concealing her in New York, where they employ her as an unconcious agent in the circulation of counterfeit money. She escapes from the wardship of an old misshapen hag, Moll, and is brought back to her home at Chestnut Wood; where, however, she is still subject to occasional manifestations from the same source.

Opinions of the Press.

"One of the pleasantest characters in the book is Jerry Goldsmith, a Yankee Calot Quotem, ready to turn his hand to any thing, and more profuse in promise than performance."—*Churchman.*

"One who has read it from *preface* to *finis*, pronounces it delightful; and hence our praise. She says there are spots that those who have tears can cry over, but never so sad that the tears need scald much."—*N. Y. Daily Times.*

"We commend to men, women, and even children, a perusal of 'Chestnut Wood.'" *Lawrence Sentinel.*

"This work will be read. It has all the elements of a successful book, viz: originality, interest, power, and strong characterization."—*Berks County Press.*

"It will please from its truthfulness to nature, and from the effect it will leave on the mind of the reader."—*Hartford Courant.*

"Its plot is well developed, is ingenious, but not too intricate, and is managed throughout with the skill of a master."—*Palladium.*

"The characters are very well and forcibly drawn, particularly the 'cute Yankee, Jerry Goldsmith."—*Mobile Adv.*

APPLETONS'

MINIATURE CLASSICAL LIBRARY

OF CHOICE EUROPEAN AUTHORS.

This unique Library comprises many of the Gems of English Standard Writers. They are works, suitable in point of size for the pocket, and are sold at extremely low prices. Each volume is illustrated with a neat steel frontispiece, and bound in a neat and tasteful form.

BOND (REV. ROBERT), GOLDEN MAXIMS; or, A Thought for every Day in the Year, Devotional and Practical. 1 vol. 32mo. 31 cents.

"The little and short sayings of wise and excellent men, are of great value, like the dust of gold, or the least sparks of the diamond."—*Tillotson.*

CLARKE'S SCRIPTURE PROMISES. In this edition every passage of Scripture has been compared and verified. 1 vol. 32mo. 38 cents.

"This volume is like an arranged museum of gems and precious stones, and pearls of inestimable value."—*Dr. Wardlaw.*

ELIZABETH; OR, THE EXILES OF SIBERIA. By Madame Cottin. 31 cts.

This little book numbers its readers by the hundred thousand.

GOLDSMITH'S VICAR OF WAKEFIELD. 1 vol. 32mo. 38 cents.

GOLDSMITH'S ESSAYS ON VARIOUS SUBJECTS. 38 cents.

Goldsmith, both in prose and verse, was one of the most delightful writers in the English language.

GEMS FROM AMERICAN POETS, consisting of Selections from the Poets of America. 1 vol 38 cents.

"This little volume will be treasured by the true lovers of poetry."

HANNAH MORE'S PRIVATE PIETY. 2 vols. 75 cents.

HANNAH MORE'S PRIVATE DEVOTION. 1 vol. 32mo. 32 cents.

"Any person who can peruse these volumes of this highly esteemed authoress, without additional wisdom or benefit, must be either superior or inferior to our common humanity."

HEMANS' SONGS OF THE AFFECTIONS. 1 vol. 32mo. 31 cts.

"No one can fail to admire this beautiful poetess, and these little poems are culled from her choicest gems."

HOFFMAN'S LOVES' CALENDAR, LAYS OF THE HUDSON, and other POEMS. 1 vol 38 cts.

"Hoffman is one of our most charming American poets, and this little volume should be in the possession of every appreciator of good poetry."

JOHNSON'S HISTORY OF RASSELAS, PRINCE OF ABYSSINIA. A Tale. 1 vol. 38 cts.

The fund of thinking which this work contains, is such, that almost every sentence of it may furnish a subject of long meditation.

MANUAL OF MATRIMONY AND CONNUBIAL COMPANION; gathered together for the safety of the single and trial of the wedded. By a Bachelor. 31 cts.

No person contemplating marriage should fail to secure a copy of this book.

MOORE'S LALLAH ROOKH. An Oriental Romance. 1 vol. 32mo. 38 cts.

This exquisite poem has long been the admiration of readers of all classes.

MOORE'S IRISH MELODIES. 1 vol. 32mo. 38 cts.

These celebrated melodies are too well known and appreciated to need much eulogy, they breathe throughout a spirit of nationality in the language which Moore alone could give.

PAUL AND VIRGINIA. From the French of Bernardin de St. Pierre. 31 cts.

The translation of this beautifully written story has maintained a reputation which is hardly surpassed by any other.

POLLOK'S COURSE OF TIME. 1 vol. 38 cts.

Few modern poems exist which at once attained such acceptance and celebrity as Pollok's Course of Time.

PURE GOLD FROM THE RIVERS OF WISDOM. 1 vol. 32mo. 31 cents.

A collection of short extracts on religious subjects from the older writers, Bishop Hall, Sherlock, Barrow, Paley, Jeremy Taylor, &c.

THOMSON'S SEASONS. One volume, 32mo. 38 cents.

Place the Seasons in any light, and the poem appears faultless; the episodes are delicious stories, the descriptions so accurate as to bear the closest test; the versification richly harmonious, yet always in perfect keeping with the subject.

TOKEN OF THE HEART, 31 cents; TOKEN OF AFFECTION, 31 cts.; TOKEN OF REMEMBRANCE, 31 cts.; TOKEN OF FRIENDSHIP, 31 cts.; TOKEN OF LOVE, 31 cts.

These little manuals are full of the excellence of poetry, each volume containing subjects appropriate to the title.

USEFUL LETTER WRITER. Comprising a succinct Treatise on the Epistolary Art, and Forms of Letters for the Ordinary Occasions of Life. 1 vol. 38 cts.

This little volume is admitted, by competent judges, to be the best manual of Epistolary Correspondence ever published.

WILSON'S SACRA PRIVATA. The Private Meditations and Prayers of the Right Rev. Bishop Wilson. One Volume 32mo. 38 cents.

This well-known work is probably the best devotional treatise in the language, and it now appears in a dress worthy of its character.

YOUNG'S NIGHT THOUGHTS. 1 vol. 32mo. 38 cts.

"In his 'Night Thoughts,' Young exhibits entire originality of style, elevation of sentiment, grandeur of diction, and beauty of imagery, accompanied with an extensive knowledge of men and things, and a profound acquaintance with the feelings of the human heart."

www.ingramcontent.com/pod-product-compliance
Lightning Source LLC
LaVergne TN
LVHW020236110826
845151LV00003B/933